CALIFORNIA NATURAL HISTORY GUIDES

INTRODUCTION TO AIR
IN CALIFORNIA

California Natural History Guides

Phyllis M. Faber and Bruce M. Pavlik, General Editors

AIR IN CALIFORNIA

David Carle

UNIVERSITY OF CALIFORNIA PRESS
Berkeley   Los Angeles   London

This book is dedicated to my family in California: sisters and brother, nephews and nieces, aunts, uncles, and cousins. May the air we share be clean and clear.

University of California Press, one of the most distinguished university presses in the United States, enriches lives around the world by advancing scholarship in the humanities, social sciences, and natural sciences. Its activities are supported by the UC Press Foundation and by philanthropic contributions from individuals and institutions. For more information, visit www.ucpress.edu.

California Natural History Guide Series No. 87

University of California Press
Berkeley and Los Angeles, California

University of California Press, Ltd.
London, England

© 2006 by the Regents of the University of California

Library of Congress Cataloging-in-Publication Data

Carle, David, 1950–
    Introduction to air in California / David Carle.
      p.   cm. — (California natural history guides ; 87)
    Includes bibliographical references and index.
      ISBN-13, 978-0-520-24531-0 (cloth : alk. paper); ISBN-10, 0-520-24531-8 (cloth : alk. paper)
      ISBN-13, 978-0-520-24748-2 (pbk : alk. paper); ISBN-10, 0-520-24748-5 (pbk. : alk. paper)
      1. Air quality—California. 2. Air—Pollution—California. 3. Atmosphere.
I. Title. II. Series.

TD883.145.C2C37 2006
363.739'209794—dc22   2005034484

Manufactured in China
10   09   08   07   06
10   9   8   7   6   5   4   3   2   1

The paper used in this publication meets the minimum requirements of ANSI/NISO Z39.48–1992 (R 1997) *(Permanence of Paper).* ∞

Cover : Wind turbines at Altamont Pass. Photograph by Warren Gretz.

The publisher gratefully acknowledges the generous
contributions to this book provided by

the Gordon and Betty Moore Fund
in Environmental Studies
and
the General Endowment Fund of the
University of California Press Foundation.

CONTENTS

ACKNOWLEDGMENTS

A book like this draws from many sources, and the writing could never be accomplished (certainly not as well) without the help of many readers and friends. I am grateful for the expert criticism of Connie Millar, research scientist with the U.S. Department of Agriculture, particularly on climate variability topics, but also for her careful reading of the entire manuscript. Also to Robert Fovell, of the UCLA Atmospheric and Oceanic Sciences Department, and Steve Ziman, Senior Staff Scientist on Air Issues and Technology with Chevron Texaco. Once again, my thanks go to Sally Gaines and Rick Kattelmann for providing feedback and suggestions on the first drafts. Several of Rick's excellent photographs also appear in this book. Other readers were Elin Ljumg, Ryan Carle, and my first reader and editor, my wife, Janet. John Werminski helped me trace several quotation sources; I envy his personal library and deep knowledge about early California.

Some of the best resources for the general public regarding air topics can be found online in web pages provided by various federal agencies at www.airnow.gov and by the California Air Resources Board at www.arb.ca.gov, where links are provided to most of the air districts in this state. An extensive bibliography accompanies this text, but several especially helpful books were: Richard P. Turco's *Earth Under Siege: From Air Pollution to Global Change* (Oxford University Press, 1997); Richard C. J. Somerville's *The Forgiving Air: Understanding Environmental Change* (University of California Press, 1998); *The Weather Book,* by Jack Williams (USA Today, 1997), which illustrated the fundamentals about U.S. weather; and *Taken for a Ride: Detroit's Big Three and the Politics of Pollution,* by Jack Doyle (Four Walls Eight Windows, 2000).

Special thanks go to Julie Popkin, who has been my literary

agent now for more than a decade. Also to Phyllis Faber and Doris Kretschmer, editors at UC Press for this and for my earlier book, *Introduction to Water in California*. Though Doris has retired, she and Phyllis launched this new subseries within the California Natural History Guides that explores human influences on California's environment.

INTRODUCTION
The Nature of California Air

It ought to be easy to take California's air for granted. Every day we take about 20,000 breaths, moving more than 3,400 gallons of air in and out, but breathing is an unconscious act most of the time, from the moment of our birth. The balance between photosynthesis and respiration is a fundamental basis for life, yet the air connection between plants and animals remains invisible to many people. And California's relatively undemanding climate makes it even easier to remain oblivious to air as a fundamental fact of life. But despite all this, most Californians actually *do* consider their air a compelling topic of conversation and concern.

Ancient Greeks saw air as one of nature's basic elements, along with earth, fire, and water. Californians see it. . . . Ah, perhaps those words offer clarification: Californians *see* too much of something that ought to be less visible. They also *feel* effects from breathing that air, which too often brings the routine act of respiration to their attention.

Growing up in the southern California coastal basin, I became accustomed to hearing weather forecasts that devoted as much time to inversions and ozone levels as they did to precipitation and temperature forecasts. I was a high school distance runner, and on many evenings it was impossible for me to take a deep breath without triggering a cough from an irritated throat. Although that seemed like a fact of life to be endured, sucking in daily doses of pollutants inevitably kindled *my* personal interest in air.

What is air? Where does it come from? How does air, in its different manifestations around California, influence us? Why do most people react positively to mountain air or sea air? How does desert air differ from the air of the Central Valley? In what ways have we disturbed natural atmospheric processes? How is pollution impacting other organisms that share our landscape? And how is global climate change affecting Californians?

This natural history guide touches on daily weather, seasonal climate, characteristic winds, and sky phenomena. "Air" includes the gases of the atmosphere, but also the aspects of air that influence all of our senses—air's taste, smell, feel, and visible nature. This book further explores our history of air quality management, impacts of air pollution on humans and the broader environment, and technological and individual measures needed to address challenges to air quality.

"The purity of the air in Los Angeles is remarkable. The air . . . gives a stimulus and vital force which only an atmosphere so pure can ever communicate" (Truman 1874, 33–34). When Benjamin Truman wrote that description, fewer than 11,000 people lived in the city of Los Angeles and the state's total population was under 600,000.

The history of California's air quality bears a direct relationship to human numbers and, in the twentieth century and beyond, the number of miles they drive their cars. Since 1940, over 60 percent of the air pollution in this state has been associated with cars and trucks. In that year, seven million people lived in California. By 2004, population growth had boomed along for six more decades, with only a few brief slowdowns, and the populace had swelled to more than 36 million, with most of us living in southern California or the greater Bay Area and in several inland urban centers. Back in 1940 (pl. 1), Californians owned about 2.8 million motor vehicles. By 2000, the much larger population drove 23.4 million cars and trucks. Total vehicle miles traveled in those six decades jumped from 24 billion to 280 billion; with a population five times as large, we were traveling 20 times as far.

The connection between vehicles and air pollution did not immediately become obvious.

Plate 1. Covina Valley, 1941. Palm trees outline orchards, with the San Gabriel Mountains in the background. Image by Spence Air Photos.

In the summer of 1943, the state's first severe smog episodes occurred in Los Angeles. Visibility dropped to only three blocks. People complained of smarting eyes, raspy throats, and nausea. The nation was at war, so perhaps the headline in the *Los Angeles Times* on July 27, 1943, was understandable: "City Hunting for Source of 'Gas Attack.'" The "gas attack" was actually "smog," and the "enemy" attacking Los Angeles' air quality would eventually be recognized as us, each of us, who contributes pollution to the shared atmosphere.

"Smog" was a term that combined "smoke" and "fog" to describe deadly air pollution episodes in England. More than 4,000 deaths occurred in London in a 1952 episode. The English smog came from coal burning that produced sulfur dioxide and soot particles. Southern California's new air pollution problem originated from different sources and had a different chemistry. At first, many blamed one synthetic rubber manufacturing plant, but the situation did not improve when the butadiene factory was shut down. In time, Los Angeles' air pollution would be identified as a distinct type termed "photochemical smog," or sometimes just "Los Angeles–type smog."

The new governor of California in 1943 was Earl Warren. He was concerned about the unknown hazards from this new phenomenon. He recommended legislation to:

> authorize research to determine what the effect of smog is on the health of the people of the state. The reception given to that suggestion was really something. One would think I had robbed the treasury. Los Angeles was the only county in the state where smog was known to be a problem at that time, and the entire Los Angeles delegation rose in its wrath and declaimed, "This is our own local problem and we will solve it ourselves." They were cheered by the lobbyists for the oil refineries, the oil companies selling gasoline, the truckers who fouled the air with the exhaust of their diesel fumes, rubber manufacturers, garbage burners, and other elements of the smog culture. The legislators of the rest of the state were not particularly interested…and my bill went down the drain without a hearing. (Warren 1977, 229)

Air pollution is a diffuse problem, the shared fault of many emitters. It is a classic example of "the tragedy of the commons." Each of us using the common resource, the atmosphere, makes choices based on our individual benefits, yet the costs are dispersed among the entire population. The benefits of car ownership are very personal, but the pollution that comes from *our* car is a minuscule fraction of the entire smog problem. It is only because there are so many cars that they create regional air pollution. The problem seems too big for any individual to resolve.

Governments exist to address such challenges to the public welfare. Despite objections by industry, in 1945 the city of Los Angeles did establish its first air pollution control program: the Bureau of Smoke Control. The bureau first targeted emissions from stationary sources like oil wells, refineries, and factories.

Statewide authorization for air pollution control districts in every California county arrived in 1947. The federal government lagged behind. In 1955, a Federal Air Pollution Control Act was enacted that provided for research and technical assistance to better understand the causes and effects of air pollution, but it lacked any regulatory clout. California established a state Bureau of Air Sanitation within the State Department of Public Health in 1959. The Motor Vehicle Pollution Control Board was also created to test and certify emission control devices for cars sold in the state.

The Federal Motor Vehicle Act of 1960 required *only* more research to address pollution from motor vehicles. Finally, in 1964, the Federal Motor Vehicle Air Pollution Control Act provided for direct regulation by the federal government. The Department of Health, Education, and Welfare was directed to establish auto emission standards.

Yet, when California adopted auto tailpipe emission standards for hydrocarbons and carbon monoxide in 1966, they were the first standards of their kind in the nation. The next year, the California Air Resources Board was created by merging the California Motor Vehicle Pollution Control Board and the Bureau of Air Sanitation.

The 1967 Federal Air Quality Act established a framework for air quality control regions based on weather and topography. Recognizing California's leadership in this area and its unique need for more stringent controls, the act also gave the state a waiver to set and enforce its own emissions standards for new vehicles (pl. 2). This special status would become controversial, eventually, after other states began voluntarily adopting California's higher standards against the wishes of polluting industries.

California's smog problem peaked in the 1970s. Significant progress against the encroaching pollution occurred in

Plate 2. Smog fills the Los Angeles basin and the Inland Empire region in 1969, piling against the San Gabriel Mountains.

the following three decades, despite a growing population and a steady increase in miles driven. The South Coast Air Basin had no Stage 1 smog alerts for ozone in 2000, dropping from 42 in 1990, 102 in 1980, and 121 back in 1977 (fig. 1).

The state, it seemed, had turned a corner. More and more people began to believe that there might not be a smog problem much longer. Progress, however, was not yet victory. We had to be careful not to declare, "Mission accomplished," before the task was completed. Unfortunately, most southern Californians and residents of the Central Valley still breathed the worst air in the nation. Although incidents of the most severe ozone violations were way down, according to the California Air Resource Board, levels of ozone and particulate pollution continued to impair health and were linked to too many early deaths.

Air could not yet be taken for granted in California. In 2003, the Public Policy Institute of California's "Special Survey on Californians and the Environment" found that most of the

Figure 1. Cartoon in the *Los Angeles Daily News*.

state's residents viewed air pollution as a serious health hazard and the most critical environmental matter facing the state. Central Valley residents were most likely to regard air pollution as the state's top environmental issue, even more than those in Los Angeles County. Most of those surveyed (65 percent) indicated their willingness to support stricter air pollution regulations on new cars, trucks, and sport utility vehicles, even if such standards would increase new-car prices.

After growing up in southern California, I embarked on a career as a state park ranger, which allowed me to experience air around this state. The coast's salty sea breezes were far different from the Central Valley's sultry summer heat. The haze of both the coast and the valley made sharp contrasts to the deep blue of Sierra Nevada skies. Desert air felt and tasted different than the atmosphere beneath redwood rainforests. Years at Mono Lake revealed connections between local dust storms and distant urban growth. Legal wrangling over water diversions that endangered that inland sea was resolved partly because of air quality requirements. The level to which the lake will be refilled was set by determining how much dry alkaline lake bed needed to be re-covered to control un-healthy dust storms and to comply with air standards. That local air should be impacted by the growth of a thirsty metropolis 350 miles away was one sign of many atmospheric connections across this large state.

This book is a natural history guide, but one that recognizes the overwhelming role of humanity in the story of air in California. "The Thin Blanket" explores the nature of the atmosphere and the chemical essence of air. "Air Apparent" is a field guide to the sky, explaining color and light, clouds and wind, and the nature of flight. "California Air Basins" describes regional differences of topography and weather, and the character of air in the state's 15 designated air basins. "Footprints in the Air" covers ways we have altered the atmosphere of California, damaging ourselves and other life in the state. "Sharing Air with the Globe" addresses the challenge of global climate change in California. Finally, "Breathing Easy" explores how we have addressed all of these challenges and suggests further actions each of us can take to improve the state's air quality and be better stewards of the global climate.

THE THIN BLANKET
The Atmosphere

For the first time in my life, I saw the horizon as a curved line. It was accentuated by a thin seam of dark blue light, our atmosphere. Obviously this was not the "ocean" of air I had been told it was so many times in my life. I was terrified by its fragile appearance.

—ULF MERBOLD, GERMAN ASTRONAUT ON SPACE SHUTTLE IN 1988
(SAGAN 1997, 173)

AS THE SPACE SHUTTLE orbits over California, the state's distinctive valleys, coastline, and the spine of the Sierra Nevada may be visible. The curvature of the Earth is also obvious along the horizon. A thin, intensely blue line traces that curve against the blackness of space, marking the atmosphere (pl. 3). The astronaut's perspective reveals just how thin the air layer is that envelops the globe, analogous in thickness to the skin of an apple. What, exactly, is this thin blanket?

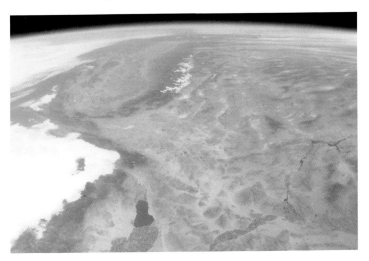

Plate 3. The thin blue line of the lower atmosphere arcs over the Central Valley and the Sierra Nevada, as seen from the space shuttle.

Layers of the Blanket

Boundaries between layers of the atmosphere are identified by differences in the way temperature changes with altitude (fig. 2). The zone where we live and where weather happens is called the troposphere. Here the air is most dense. The troposphere includes the first seven miles up from the ground in middle latitudes (at the equator it can extend up over 11

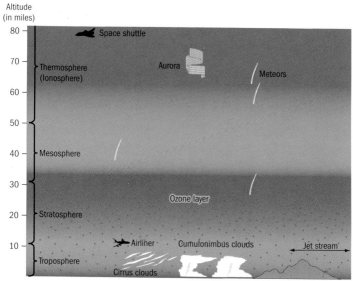

Figure 2. Atmospheric layers.

miles). Temperatures decrease as altitude increases in this level, except where there are local inversions (see "Temperature Upside Down," later in this section).

The stratosphere, where temperatures get warmer with height, extends from the troposphere to about 35 miles above the Earth's surface. The lowest portion of the stratosphere sits above 80 percent of the global air mass. Temperatures climb in the stratosphere because some of the sun's energy is captured here by ozone. The ozone layer shields life on the surface of the Earth from the harmful effects of ultraviolet sunlight.

The upper atmosphere is formed by the mesosphere and thermosphere. The mesosphere, above the stratosphere, extends up to about 50 miles above the ground. Temperatures there fall once again, to −130 degrees F. No conventional aircraft flies this high, but some large helium balloons are sent into this layer by researchers. The mesosphere is where almost all small meteors burn up; we see them at night as "shooting stars."

Outermost is the thermosphere, which extends out several hundred miles above the Earth's surface. Here is where the space shuttle orbits and where auroras, the "northern lights," occur. Air density is extremely thin in this region, yet temperatures become extremely hot (sometimes thousands of degrees, depending on solar activity). The thermosphere is ionized, containing charged particles that facilitate long-distance radio communication. Signals from the ground can hit this "ionosphere" (the ionized region that actually includes some of the mesosphere) and bounce back to distant ground stations around the globe.

All these layers make up the "thin blanket." As distant as the upper layers seem, even the air found far overhead interacts with the oceans, land, and life on the surface.

The Elegant Balance: Photosynthesis and Respiration

Air is the medium for a global respiratory system. An essential player in life, it provides carbon dioxide for plant photosynthesis and oxygen to support the respiration of both plants and animals. Some of the air we breathe is manufactured by plants and bacteria within California, but the majority is imported from oceans and from distant continents by global winds and air masses.

As we watch squash plants emerge from tiny seeds and grow into sprawling garden vines, or tree saplings gradually become tall enough to shade a house, it is amazing to realize that all such growth was accomplished by pulling an invisible gas, carbon dioxide, out of the air. Photosynthesis (literally, "making with light") allows green plants to use solar energy to transform carbon dioxide and water into complex carbon molecules. The recipe incorporates six carbon dioxide molecules with six molecules of water to build glucose. That product is then used by plants to ultimately build stems, leaves, flowers, and roots. Oxygen gas is released to the atmosphere during photosynthesis. Though a waste product for plants, oxygen is, of course, essential to us and most other life in California. The summary of this photosynthesis reaction is:

carbon dioxide + water + sunlight (energy) → sugars + oxygen

We cannot usually see the gases, carbon dioxide and oxygen, coming and going from plant leaves, except when aquatic plants are submerged. When the sun shines on algae in shallow freshwater springs, the water surface may become covered with tiny bubbles—visible evidence of photosynthesis under way (pl. 4). Bend over close to look; take a hit from the "oxygen bar."

Respiration reverses the photosynthesis equation by using oxygen to burn complex carbon molecules and release carbon dioxide, water, and energy. You are reading these words and turning the pages of the book utilizing energy from respiration. You breathe in oxygen, exhale carbon dioxide, and par-

Plate 4. Algae giving off oxygen bubbles.

ticipate in the elegant balance. So it is life—on land, in the oceans, and passing through the air itself—that generates and maintains the atmospheric gases that we so easily take for granted.

Why Is There Air? Evolution of the Atmosphere

Our atmosphere and life probably coevolved (fig. 3). The current oxygen-rich atmosphere could exist only after photosynthesis appeared. Following a long period with only anaerobic ("without oxygen") bacteria, the first green plants developed. Oxygen, the product of photosynthesis, did not transform the atmosphere overnight. If all the complex carbon molecules were soon converted back to carbon dioxide, that closed cycle would ensure that no surplus oxygen accumulated. Yet an oxygen-rich atmosphere did ultimately appear because some carbon compounds settled to the bottom of the seas, were buried in sedimentary layers, and never decomposed. Instead they became soil, rock, or, transformed by heat and pressure, today's fossil fuels (coal, oil, and natural gas). Surplus oxygen

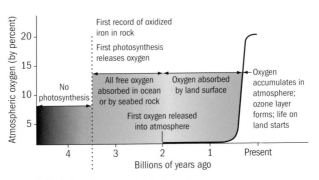

Figure 3. Evolution of an oxygen-rich atmosphere.

would, for a time, have been completely caught up in reactions with iron in the oceans (the oxidation that produces rust). Finally, after ferrous compounds were saturated, the oxygen level could build toward the modern equilibrium concentration, 21 percent of atmospheric gases.

This oxygen concentration itself is balanced elegantly with life. Scientist James Lovelock has speculated that oxygen concentrations above 25 percent would make *everything* flammable, even damp wood, but if the level fell below 15 percent, not even the driest twigs would burn. If oxygen levels were too low, there would not be enough to support metabolic respiration, but if too high, the reactive nature of oxygen would destroy living cells. "We are so accustomed to think of oxygen as life-saving and essential that we ignore its potent toxicity" (Lovelock 1995, 122).

What a Difference an O Makes

Another molecular form of oxygen is ozone (O_3), which is even more reactive than oxygen gas. Ozone is a trace gas, even within the "ozone layer" of the stratosphere. If all the ozone in the atmosphere could be gathered at sea level pressure and temperature, it would form a band only about one-tenth of an inch thick. Concentrations of oxygen gas far exceed that of ozone, yet the trace ozone amounts play vital roles, both "good" and "bad" from our perspective. Ozone near the ground is the major pollutant and health hazard in photochemical smog (see Footprints in the Air). In the stratosphere, ozone does us an essential favor by absorbing ultraviolet rays from sunlight. Until the ozone shield formed, after the development of an oxygen atmosphere, life was necessarily confined to water or other places protected from direct sunlight. The high-energy ultraviolet rays would have been toxic to early life exposed on land.

Some ultraviolet light does still penetrate the ozone shield and reach the surface. We need enough skin exposure to ultraviolet rays to form vitamin D, but those rays are why we are warned to slather on sunscreen and avoid sunburns, or risk skin cancer and eye damage.

Ultraviolet energy absorption in the stratosphere explains why temperatures are higher there and also explains the source of the ozone itself. Some oxygen molecules are broken apart by ultraviolet light; the O atoms formed combine with other oxygen molecules to make ozone.

The ozone then can itself absorb ultraviolet light, converting energy to heat, but also reversing the earlier reaction, giving back an O atom and an O_2 molecule. Equilibrium is maintained as sunlight drives both ozone formation and its destruction. Life on the surface, including our lives, benefits from that process. (Human impacts to the ozone layer are discussed in Sharing Air with the Globe.)

The Essence of Air

The atmosphere participates in nutrient cycles where nitrogen, carbon, and water change form, pass through living matter, and return to the environment (table 1). California's air connects its life forms while replenishing and maintaining an atmospheric balance.

The Nitrogen Cycle

Nitrogen (N_2) is, by far, the dominant gas in air; it makes up about 78 percent of each breath you take in dry air. Yet nitrogen gets overlooked by people who focus on what seems to matter most, the oxygen that we keep inhaling to stay alive. Nitrogen is just as essential, but in different ways. It is a key building block in proteins, enzymes, vitamins, and DNA. We cannot directly make use of it from the air. The gas is a very stable molecule; it takes energy to "fix" nitrogen so it can then be passed along in food chains. Animals can only take in nitrogen by eating plants or other animals that have eaten plants. Even plants only use nitrogen that has first been

TABLE 1.   Ten Most Common Components of Dry Air

	Volume in Air (%)
MOST ABUNDANT GASES	
Nitrogen	78.08
Oxygen	20.94
OTHER TRACE GASES	
Argon	.934
Carbon dioxide	.037
Neon	.00182
Helium	.00052
Methane	.00015
Krypton	.00011
Hydrogen	.00005
Nitrous oxide	.00005

The total for nitrogen and oxygen is 99.02 percent. The volume of carbon dioxide varies but is increasing. "Dry" air always contains some water vapor, but in varying amounts that cause the other gas percentages to adjust accordingly.

"fixed" by bacteria, by lightning, or during combustion (whether in wildland fires or in manmade engines) (fig. 4).

Combustion causes nitrogen to combine with oxygen in the air to produce nitrates (NO_3^-). These water-soluble compounds are delivered to the ground by rain as nitric acid. Once moved from the air to the soil, the nitrogen in nitrates is available to plants for building complex molecules.

Ammonia (NH_3) is the other key nitrogen-containing nutrient for plants. Nitrogen-fixing bacteria make ammonia by combining nitrogen from the air with hydrogen. Some nitrogen fixers are found in the soil, but others coexist with specialized plants, housed inside nodules on roots of legumes, such as peas, beans, and clovers. Still other bacteria can further convert ammonia to nitrates.

Nitrogen is often a limiting nutrient for plants. Levels can drop because nitrogen nutrients are water-soluble and wash away. Farmers sometimes plant legumes in rotation with

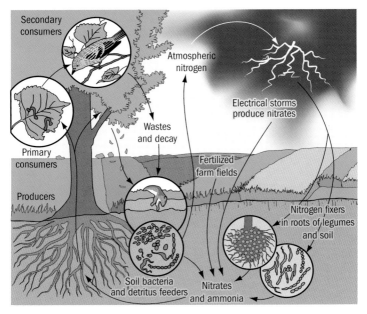

Figure 4. The nitrogen cycle.

other crops to replenish nitrogen in the soil. They also apply both ammonia and nitrate as fertilizers. When fertilizers enter runoff to streams or when nitrous oxide in air pollution falls onto lakes and streams, excess nitrogen is deposited into water systems. If the water ecosystems are not adapted to such levels of nitrogen, changes happen. Lake Tahoe is renowned for crystal-clear water almost devoid of algae. Nitrogen deposits from development and air pollution in the Tahoe Basin have fertilized unnatural algae growth (a process called eutrophication) and diminished the lake's clarity.

The cycling of nitrogen becomes complete as animals excrete urea or when amino acids are broken down during decomposition. Denitrifying bacteria (common in wetland muds and estuaries) convert amino acids, ultimately, back to nitrogen gas.

The Carbon Cycle

Carbon is the atom that is the framework for all organic molecules. It enters ecosystems from the atmosphere when carbon dioxide is captured in photosynthesis (fig. 5). A trace gas, carbon dioxide (CO_2) makes up a very tiny portion of the atmosphere, only 37-thousandths of 1 percent (.037%). The amounts measured atop Mauna Loa, in Hawaii, vary in a natural annual cycle. In spring and summer, when Northern Hemisphere plants are photosynthesizing, the atmosphere loses carbon dioxide (about a 3 percent seasonal drop). In fall and winter, with less photosynthesis happening, the carbon dioxide level goes up. This seasonal rise and fall has been called "the signature of the biosphere breathing" (Somerville 1998, 40).

About 10 percent of the carbon dioxide in the atmosphere is cycled by photosynthesis and respiration each year. There is uncertainty about whether ocean or land plants carry out the most photosynthesis. Estimates range from half occurring in the oceans, to one-third in the ocean and two-thirds on land. On land, fast-growing rainforests, including the temperate rainforests of California's coast redwoods (*Sequoia sempervirens*), pull particularly large amounts of carbon from the atmosphere into long-term storage in trees. Soils store carbon too, holding four times as much as the atmosphere and three times more than the world's trees.

Carbon dioxide levels have been increasing by four-tenths of 1 percent per year in recent years. There are natural episodic events, like volcanic eruptions, that temporarily inject higher levels of CO_2 into the atmosphere. However, the recent rise—to levels not equaled for perhaps a half-million years—is almost certainly due to our burning of fossil fuels. As CO_2 increases, oxygen levels slowly decline. We face no danger of running out of oxygen, however, because CO_2 is only a trace gas in the air. Even if we burned all the earth's fossil fuel, all the trees, and all the organic matter in soils, only a small percentage of atmospheric oxygen would be consumed.

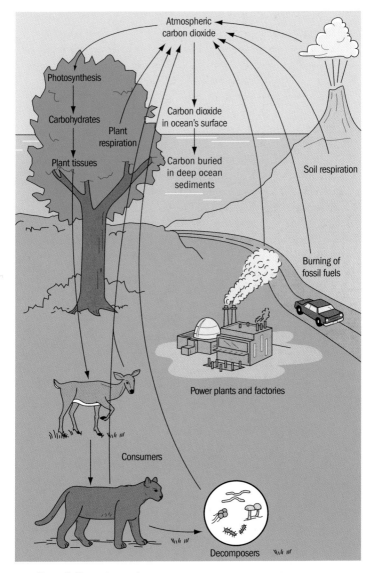

Figure 5. The carbon cycle.

The concern about CO_2 has far more to do with its role as a greenhouse gas.

The Essential "Greenhouse Effect"

The analogy of an atmospheric process to a greenhouse refers to the way glass windows can allow sunlight to enter, while keeping infrared heat waves from exiting (fig. 6). The atmosphere, similarly, is transparent to most sunlight, letting it pass through the air. When heat radiates back from the surface, though, some air molecules absorb the longer infrared wavelengths. Without this natural greenhouse effect, the Earth would have an average temperature of only about 5 degrees F. In that case, most familiar life forms could not exist. Temperatures would be closer to those on the Moon, where there is no air to create a greenhouse effect. Heat absorbed by the lunar

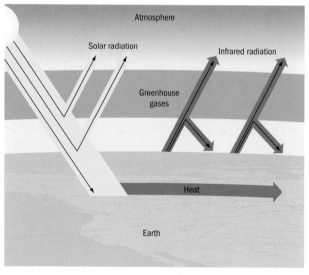

Figure 6. The greenhouse effect. (*Note:* Greenhouse gases mix throughout the atmosphere. They are depicted here as a layer to emphasize how they trap heat.)

surface simply radiates away. Though the Moon is about the same distance from the sun as the Earth is, it is much colder.

It is the *enhanced* greenhouse effect that currently causes enormous concerns. Human burning has elevated carbon dioxide levels by releasing carbon stored in fossil fuels. Since 1800, coinciding with the industrial revolution, carbon dioxide levels have risen from about 270 parts per million (ppm) to over 370 ppm (fig. 7). The level will probably climb past 540 ppm, doubling the preindustrial amounts around the year 2050. If there is no change in the generation of greenhouse

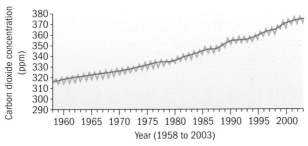

Figure 7. Carbon dioxide in the atmosphere, measured at the Mauna Loa Observatory, 1958–2003.

gases by humans, levels could climb much higher by the end of the century. Both ocean and land plants capture some of the increased amounts of carbon dioxide; growth can even be "fertilized" by high carbon dioxide levels. However, the rising rate of CO_2 production overwhelms vegetation's ability to maintain a balance. (Higher global temperatures and other effects on climate will be covered in Sharing Air with the Globe, along with measures to address these problems.)

Other greenhouse gases include water, methane, and nitrous oxide. Some man-made chemicals, like chlorofluorocarbons (CFCs), also act as greenhouse gases. Although the focus is often on carbon dioxide as a greenhouse gas, the whole package of gases is important. For example, methane levels have increased from agricultural sources like rice fields (fertilized with organic matter that decomposes to release the gas) and cattle, which generate the gas from both ends. Each methane molecule is much more persistent in the atmosphere than carbon dioxide and, molecule for molecule, far more potent at capturing heat.

A Dating Game: Ions and Radiation

Our planet's atmosphere is constantly being struck by cosmic rays that travel near the speed of light. They are very high-energy, charged particles blasted from supernovae beyond our solar system. Cosmic rays create isotopes, altered versions, of our atmospheric gases. Energy from cosmic rays can convert nitrogen to carbon-14, a radioactive atom whose nucleus has eight neutrons instead of the usual six. Carbon-14 is unstable. It gradually decays; half of the material will achieve a stable state in 5,730 years. That predictable half-life can be used to date organic material. Carbon-14 is incorporated into plants, along with regular carbon (carbon-12), during photosynthesis. Animals that eat plants then hold the two carbon forms in their tissues. In a living animal or plant, the ratio between the isotopes is the same as in the air. When a plant or animal dies, though, the decay of carbon-14 causes the ratio to begin changing. If a piece of old wood contains only half as much carbon-14 as living plants, its estimated age is 5,730 years.

If radiocarbon dating indicates that an organism's remains are older than a few thousand years, then the age estimate must be corrected to adjust for variations that are known to have occurred in the rates that cosmic rays strike Earth. Scientists calibrate this kind of data using annual rings in trees, correlating known ages with carbon-14 variations.

Another radioactive gas useful for atmospheric studies is radon, the product of the decay of uranium from the soil into the air. Radioactive radon gas is a health concern in some parts of California (see Footprints in the Air).

Weighty Matters: Air Pressure

Comedian Bill Cosby once joked about his girlfriend, a philosophy major in college, walking "around the house saying, 'Why is there air?' And I used to look at her, 'Well any phys. ed. major knows why there's air. There's air to blow up volleyballs, blow up basketballs!'" (Cosby, 1965).

Molecules of air inside balls exert pressure because they are constantly moving in all directions. This energetic motion also keeps air from being pulled by gravity down to the ground. However, gravity does give weight to the atmosphere. Air pressure at sea level is felt as 14.7 pounds of weight per square inch. At 9,000 feet above sea level, the pressure drops by about one-fourth, down to around 11 pounds per square inch. That is why an unopened bag of potato chips, filled at sea level, expands to nearly the bursting point in the mountains (unless the temperature has substantially dropped) (pl. 5).

As 25 million tons of air push down on each square mile of the planet, one might wonder why our bodies are not crushed. Thankfully, we have adapted to this condition. Air and fluid inside our hollow organs and vessels balance the outside pressure. We seldom become aware of the weight of air until our ears "pop" as we change altitude, when pressure is equalized on both sides of our eardrums.

Because air is a gas, it can be compressed or expanded. Under less pressure, fewer air molecules occupy a given volume. With every breath at the summit of Mount Whitney (14,495 feet above sea level), the number of oxygen molecules inhaled is about three-eighths that at sea level. Oxygen gas still makes up 21 percent of the thinner air, but there are fewer molecules of air per lungful atop the mountain.

Above 6,000 feet, some people experience high-altitude sickness. The ability of our bodies to eventually adapt to high altitude brings Olympic distance runners to train at places like Mammoth Lakes (over 8,000 feet), where their blood develops not only more red blood cells but also a more efficient gas-exchange system between the blood and body cells.

Plate 5. *Left:* A potato chip bag, expanding to near-bursting because of the low air pressure at 9,000 feet. *Right:* The bag at sea level, compressed by higher atmospheric pressure.

Commercial aircraft often cruise between 25,000 and 35,000 feet above sea level. Those altitudes would be fatal for humans if aircraft cabins were not pressurized. For most flights, the cabin pressure recreates the air pressure that is experienced at 5,000 to 8,000 feet above sea level. The short duration of flights keeps that pressure from bothering most people.

Highs and Lows: California's "Mediterranean Climate"

Weather comprises the air pressure, temperature, precipitation, wind, and cloud conditions at any given moment. The term "climate" refers to weather over time, not just averages, but also the range of variability and the extremes that characterize a location. Looking at average temperature or rainfall records never tells a complete story. Death Valley is justifiably notorious for its average summertime highs, but even more

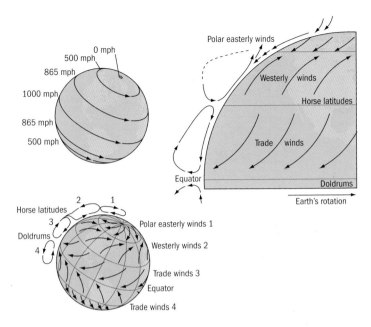

Figure 8. Global circulation of the atmosphere.

so for its terrifically hot extreme records. Average wind speeds in southern California do not reveal the importance of periodic episodes when extremely dry, hot, Santa Ana winds howl across the coastal plain.

California's climate is strongly influenced by a high-pressure region over the ocean called the Pacific High. The forces that manufacture that High begin as sunlight falls most directly onto the globe's equatorial region (fig. 8). Warm air moves up because it is lighter and less dense than cooler air. Cold air sinks. The warmed equatorial air rises and moves north and south from the equator. Around 30° latitude, in both hemispheres, the air begins to sink. The mass of subsiding air produces high pressure that raises the air temperature, lowers relative humidity, and results in clear, fair weather. In the Atlantic, this region around 30° north was named the

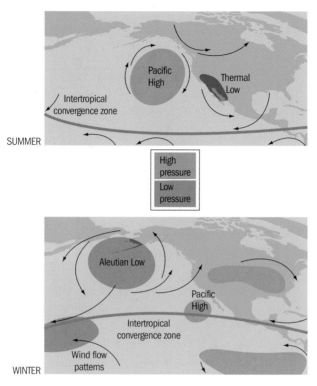

Figure 9. Positions of the Pacific High in summer and winter.

"horse latitudes" because Spanish vessels sailing to the New World became becalmed there and many horses had to be forced overboard after food ran out.

The Pacific High sits off the west coast of North America (fig. 9), centered at about the 35° north latitude line during summer (California extends from 32.5° north to 42° north). The High blocks the approach of storms from the Pacific Ocean, keeping California's summer weather dry. In winter, the High slides southward, opening a storm track for northern storms. The Aleutian Low generates storms from the Arctic that, in some winters, reach all the way into southern California. This pattern of wet winters and a summer drought is typical of Mediterranean climates. In the United States, this climate type is found only in California. Elsewhere on the globe, Mediterranean climates appear on the southwest corners of the major continental landmasses of Eurasia, South America, Australia, and South Africa, confined between latitudes 30° and 45°. In total, the five Mediterranean climate areas include only about 2 percent of the land on Earth (map 1).

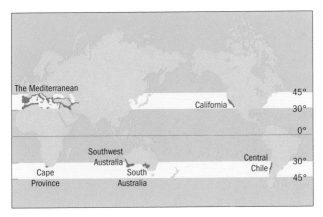

Map 1. Mediterranean-type climate regions around the world.

The Mediterranean type of climate has three variations within California (map 2). A "cool summer/cool winter" variety is found along the coasts and on the western slopes of the Sierra Nevada; a second coastal variation entails frequent summer fog; interior valleys experience a third version, marked by hotter summers and cooler winters. All these areas, however, follow the Mediterranean pattern of winter rain and summer drought.

To complete its daily rotation, our planet spins about

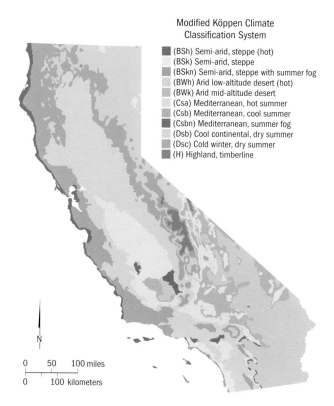

Modified Köppen Climate
Classification System

■ (BSh) Semi-arid, steppe (hot)
  (BSk) Semi-arid, steppe
  (BSkn) Semi-arid, steppe with summer fog
  (BWh) Arid low-altitude desert (hot)
  (BWk) Arid mid-altitude desert
  (Csa) Mediterranean, hot summer
  (Csb) Mediterranean, cool summer
■ (Csbn) Mediterranean, summer fog
  (Dsb) Cool continental, dry summer
  (Dsc) Cold winter, dry summer
■ (H) Highland, timberline

N

| 0 | 50 | 100 miles |
| 0 | | 100 kilometers |

Map 2. Climate zones of California.

1,000 miles per hour at the equator, about 800 miles per hour in San Francisco, and approaches zero miles per hour at the poles (fig. 8, upper left). The deflection of global winds by this rotation is called the Coriolis force. Even stronger pressure gradient forces combine with the Coriolis force to bend winds in the Northern Hemisphere in a clockwise rotation away from high-pressure zones, while spiraling the winds inward toward low-pressure centers with a counterclockwise rotation.

The Pacific High generates winds out of the west and

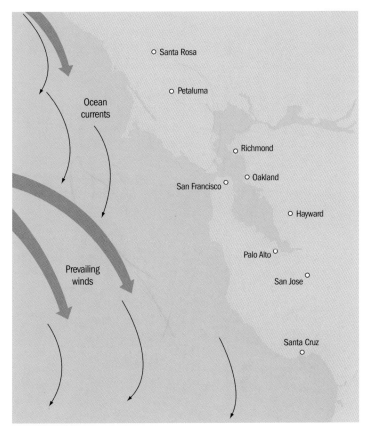

Figure 10. Coastal wind and ocean current patterns off central California.

northwest along the California coastline. Winds passing over the ocean create surface currents that also travel in the same directions. The Coriolis effect, acting on both wind and water currents as the Earth rotates eastward, turns ocean currents away from the California coast (fig. 10). To replace surface water shifting away from the coastline, water must well up from the depths. That water is colder. Northern and central California beaches are particularly renowned—or notorious—for frigid water. In northern California, swimmers at the beach might encounter water that is 50 degrees F, even in summer. The upwelling water leads to coastal fogs and low layers of stratus clouds as humid air, approaching across the warmer ocean, reaches the narrow band of cold coastal water and is chilled. Moisture in the air condenses.

Once over land, the fog and overcast layers commonly "burn off" each day, then re-form at night. Because the California coast bends to the southeast below Santa Barbara, the winds and ocean currents spiraling out of the Pacific High affect the southern part of the state less. There is usually warmer summer surf (from 57 degrees F to 65 degrees F) off southern California beaches.

Coastal marine climates are moderated by the local ocean temperatures: hot weather is "air conditioned" on summer days while cold winter days are warmed. The ocean is a very effective heat sink, storing heat during the day and then releasing it at night when the air cools off. Inland sites have "continental climates," where the daily temperature range is greater, because landmasses and topography exert a larger influence than they do on the coast.

Weather and the climate in inland portions of California are shaped by a thermal low that forms in summer as heat rises out of the Central Valley and desert basins. The rising air creates low pressure that sucks marine air inland. Onshore sea breezes form after the land heats each day. The delta breeze is a particularly strong sea breeze that moves from San Francisco Bay through the Carquinez Straits, bringing cooler air into the delta region of the Sacramento and San Joaquin Rivers and the valleys, north and south of that point.

Parts of California are also affected by a Great Basin High, which develops east of the Sierra Nevada during winter. Winds descend from the high deserts under those conditions and scour the air in the inland and coastal valleys.

El Niño

The Earth's oceans and atmosphere work together as a system. The planet is ultimately heated by the sun, but much of the warmth in the air passes first through the oceans. The upper few feet of the ocean hold as much heat as the entire atmosphere. Heat travels in warm ocean currents that release the energy and raise air temperatures when they reach cooler latitudes.

Most Californians have heard about El Niño (more accurately, the El Niño Southern Oscillation), an ocean cycle that periodically sends heavy winter rains into the state. During an El Niño event, wind and ocean currents flow west to east, and warm water collects off the west coast of North and South

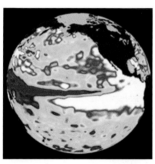

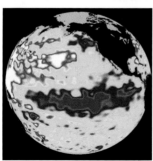

Plate 6. *Top:* El Niño conditions in the Pacific Ocean, November 10, 1997. Red and white colors indicate warmer water. *Bottom:* La Niña conditions in the Pacific, February 27, 1999. Blue and purple indicate cooler water.

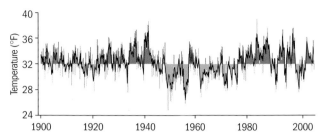

Figure 11. Pacific Decadal Oscillation (PDO), sea surface temperature variability, 1900–2004.

America. Jet streams and storm tracks overhead shift accordingly. The results vary, but strong El Niño conditions often send heavier rainstorms to northern California (with a smaller effect in the southern part of the state). In between the El Niño years, La Niña may appear, when a "cool pool" forms in the Pacific waters closest to us, as winds and ocean currents turn around to flow away from the coast. During La Niña episodes, most of California experiences drier winters. This cycling between El Niño and La Niña, every three to seven years, is a natural phenomenon that has now been traced back thousands of years (pl. 6).

There are longer ocean-atmosphere cycles at work too. Climate scientists have identified the Pacific Decadal Oscillation (PDO), a 10- to 20-year pattern that produces less spectacular effects than El Niño, but with a longer duration (fig. 11). During positive phases of the PDO, the shorter-term El Niño events get amplified. Negative phases of the decades-long cycle accentuate La Niña effects, as happened in the late 1990s, when the West Coast and the southwestern states experienced drought.

Cycles lasting much longer, on a thousand-year scale, overlay the shorter cycles. Forest Service scientist Connie Millar calls the resultant pattern a complex "symphony" that must be teased apart to understand the music of our climate (personal communication).

"It's Not the Heat,
It's the Humidity"

White patches visible near the Sierra crest amaze travelers as they bake in the heat of the Owens Valley in summer. Some find it hard to believe they can see snow when it is 95 degrees F in Lone Pine. Yet, temperatures generally drop 5.4 degrees with every 1,000 feet of elevation gain, the "lapse rate" of rising dry air as it expands and cools. On a typical summer day in Death Valley, 282 feet below sea level, the day's high temperature often reaches 115 degrees F. At the same time, at 4,000 feet in the Owens Valley, thermometers may read 94 degrees F, or 21 degrees cooler. Snow patches and glaciers persist at the high elevations of the Sierra Nevada, particularly on north-facing slopes, on peaks that may reach 14,000 feet above sea level. There, daily highs may be 75 degrees F cooler than at sea level. At night, in the dry, clear air of both the desert and the mountains, lows may drop down to 85 degrees F in Death Valley, be 64 degrees F in the Owens Valley, and plummet well below freezing at the summit of Mount Whitney.

The exact temperature changes with altitude vary, depending on many factors, including wind, humidity, and topography. One reason that travelers from out of the area have so much trouble believing that those distant white patches are really snow or ice is that many live near the coast, where the ocean moderates temperature swings and where such long-distance visibility is rarely possible.

In Death Valley, souvenir shops sell a T-shirt that proclaims: "But it's a *dry* heat." The slogan recognizes the importance of relative humidity. That measurement compares how much water vapor is in the air to the maximum amount of vapor that air could possibly hold (under given temperature and pressure conditions). Words like "sultry," "clammy," or "dry heat" express the importance humidity plays in our comfort levels with weather. We fan ourselves to move air

Plate 7. Breaking waves cast salt particles into the air.

across sweaty skin, taking advantage of water's ability to carry off heat. When the air is very humid, the cooling effects of evaporation from our skin diminish. Weather forecasters include relative humidity in their daily forecast and also in the "heat index," which measures how hot it really feels when high humidity coincides with high temperatures.

What is it that makes "sea air" appealing? Part of the answer may be explained by salt spray. Millions of tons of salts are spun into the air each year when waves burst against the world's shorelines (pl. 7). Crashing waves create foam, bursting bubbles, and aerosols, tiny particles that enter the air carrying salts. Most of the particles are large and quickly fall out of the air. Yet some are transported great distances by the wind (pl. 8). Some of the particles become nuclei around which water condenses to form clouds.

Being near ocean surf or near crashing water like Yosemite Falls seems to make people feel good. The physical explana-

Plate 8. Salt-laden mist moves inland from the surf zone.

tion is not yet clear, but the feeling may be linked to negative ions carried by small droplets generated in such places. Some research suggests that negative ions in the air can lower blood pressure, ease breathing, and improve subjective feelings of good health. By contrast, more positive ions tend to accumulate near industrial pollution and in overcrowded spaces. Positive ions may explain complaints associated with southern California's hot, dry Santa Ana winds, which get blamed for irritability, headaches, and insomnia. A connection between ions and altered levels of the mood hormone serotonin has been postulated, though not yet proven.

Not all aspects of fresh sea air may be healthy. In polluted urban coastal cities, sea salt particles may actually boost production of ozone smog by reactions with sunlight and with pollution trapped beneath marine inversions.

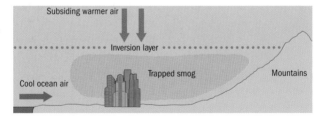

Figure 12. Inversion layer formation near the coast.

Plate 9. An inversion made visible by pollution trapped beneath it in the Central Valley, as seen from the Sierra Nevada.

Temperature Upside Down

Temperatures do not *always* decline with altitude. Inversions reverse this pattern (fig. 12). They can form when subsiding air, which gets warmer as it descends, settles above cooler air near the ground, or when sea air blows onshore and penetrates beneath air warmed by the land. Such marine inversions are typical of the Los Angeles region. Radiational inversions form differently, on cloudless nights over land where the air closest to the cold ground becomes chilled. Sometimes this pattern gets reinforced as cold air drains from slopes down onto lowlands. Pollution is trapped and accumulates in the stagnant air beneath inversions.

We cannot see "temperature" generally, but inversions often make temperature differences obvious to our eyes (pl. 9). Many other air phenomena are apparent to our sight and the rest of our senses.

AIR APPARENT
A Field Guide to the Sky

The sky just is, whether we admit it or not, whether we look at it or not, whether we love it or whether we hate it. It is quiet and big and there. If you don't understand it, the sky is a very mysterious thing, isn't it? It's always moving, but it's never gone.

—RICHARD BACH (1974, 295)

AIR AFFECTS OUR SENSES at every moment, but not always at a conscious level. We perceive it as the source of our daily weather; the medium for clouds and rain, wind and temperature. "How far it has come, and how far it has to go!" John Muir wrote:

> How many faces it has fanned, singing, skimming the levels of the sea; floating, sustaining the wide-winged gulls and albatrosses; searching the intricacies of the woods, taking up and carrying their fragrances to every living creature. Now stooping low, visiting the humblest flower, trying the temper of every leaf, tuning them, fondling and caressing them, stirring them in lusty exercise, carrying pollen from tree to tree, . . . playing on every needle, on every mountain spire, on all the landscape as on a harp. (Muir 1979 [1938], 98–99)

Less obviously, air also transmits sounds, odors, and light. In the high country, the pungent odor of pine trees, the deepening blue color of the sky, and long-distance vistas through clear mountain air can be agreeable contrasts to the haze of California's urban lowlands. Approaching the coast from inland, we may detect the first smell of salt air long before the ocean itself comes into view or before the surf can be heard smashing on shore, sending salt spray off into the sky. Air is the wind beneath the wings of flying insects, birds, and mammals (both bats and humans). It delivers pollen to fertilize flowers and disperses many of the resulting seeds. It also disperses smoke and pollutants or, at other times, clings to those particles and chemicals, trapping them in stagnant basins.

Why Is the Sky Blue?

Sunlight contains a range of frequencies of visible and invisible light. White light broken apart by prisms displays all the wavelengths within visible light, each seen as a different color. Red light has the longest wavelength in the visible spectrum;

blue and violet lights have much shorter wavelengths. Not visible are ultraviolet rays, at the high-energy and high-frequency end of the light spectrum, and infrared at the other extreme, which we perceive as heat.

Leonardo da Vinci proposed one answer to the familiar question "Why is the sky blue?" He wrote: "I say that the blue which is seen in the atmosphere is not its own color, but is caused by the heated moisture having evaporated into the most minute and imperceptible particles, which the beams of the solar rays attract and cause to seem luminous against the deep intense darkness . . . above them" (da Vinci 1956 [1508], 399). As the artist and early scientist correctly surmised, the sky itself is not full of blue-tinted particles. Instead, light waves are deflected by air molecules (not simply "heated moisture," however), and repeated collisions scatter light in all directions. We see a blue sky because shorter wavelengths are scattered more than longer ones (table 2). Blue is scattered twice as effectively as green light and about four times more than red light. Violet is actually scattered the most, but our eyes are not very sensitive to that range; we perceive blue better. Without atmospheric gases and their abilities to scatter sunlight, our sky would be as black as outer space, where light passes through but is not made visible by encounters with air molecules.

Mountain skies often appear deeper blue than skies in val-

TABLE 2. Coloring Effects Due to Scattering and Absorption of Different Light Wavelengths

	Air	Clouds	Haze
Blue light	Strong scattering	Strong scattering	Strong scattering
Green light	Moderate scattering	Strong scattering	Moderate scattering
Red light	Weak scattering	Strong scattering	Moderate scattering
Color effect:	Blue	White	Gray

Source: Modified from Turco (1997, 63).

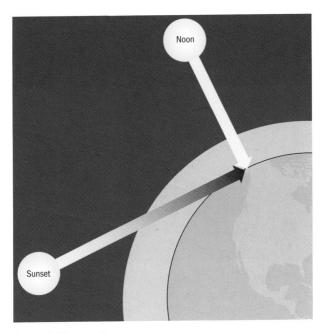

Figure 13. Why sunsets are red.

leys or along the coast. Deep blue skies are dry and clean, free of smoke, dust, and haze. Water vapor and haze particles turn the sky whitish or gray because they scatter more light across the whole range of wavelengths. Clouds also appear white due to the strong scattering of light by water vapor and ice crystals.

Why do sunsets and sunrises appear in shades of red and yellow? At those times, light is traveling through much more of the atmosphere before reaching our eyes. When the sun is just above the horizon, its light must pass through 20 times more atmosphere than when the sun shines directly overhead. Only the longest wavelengths, the yellows, oranges, and reds, penetrate that entire distance (fig. 13, pl. 10).

Muir described a special sunset phenomenon in the Sierra Nevada known as alpenglow:

Plate 10. Sunset colors.

Now came the solemn, silent evening. Long, blue, spiky shadows crept out across the snow-fields, while a rosy glow, at first scarce discernible, gradually deepened and suffused every mountain-top, flushing the glaciers and the harsh crags above them. This was the alpenglow, to me one of the most impressive of all the terrestrial manifestations of God. At the touch of this divine light, the mountains seemed to kindle to a rapt, religious consciousness, and stood hushed and waiting like devout worshippers. Just before the alpenglow began to fade, two crimson clouds came streaming across the summit like wings of flame, rendering the sublime scene yet more impressive; then came darkness and the stars. (Muir 1961 [1894], 45)

The warm tint of alpenglow appears during twilight after the original sunset colors on mountain summits have faded. Despite the sun being below the horizon and the mountain peaks passing into shadow, a re-illumination occurs as scattered light rays from the west are refracted downward.

Plate 11. A double rainbow.

The familiar arcs of rainbows display the visible spectrum normally hiding in "white light." To see a rainbow, you must have the sun behind you and be looking toward raindrops struck by sunlight. Light is bent (refracted) as it passes through the surface of a raindrop, is reflected off the back of the drop, and bends again as it exits the drop. The result is light coming back toward the observer at about a 42-degree angle from the observer's shadow. Many individual raindrops produce the illusion of a full "bow." Side-by-side observers see different rainbows, formed by the different raindrops at the correct angles. Refracted light separates into bands of color because each light frequency bends a slightly different amount. Red shows up on the outer edge of the bow, then orange, yellow, green, blue, and finally violet along the inner edge.

Double rainbows can also form. The color sequence is reversed in the secondary rainbow, which is fainter because light rays are reflected twice inside the surface of the drop.

The second "bounce" off the back surface, before the light re-emerges, bends rays even more, so that another rainbow forms at about a 52-degree angle to the observer (pl. 11).

Halos are circular rainbows that may form high overhead when sunlight is refracted by ice crystals in cirrus clouds.

How to Find Cloud 9

> Clouds . . . mark the places where water vapor can no longer hide in the sky.
>
> — LOUISE B. YOUNG (1977, 37)

All air contains some water vapor, though the amounts vary. The higher the air temperature, the more water vapor air can hold. When the temperature drops, an air mass can reach its dew point, where it can no longer contain all its water vapor; some condenses into the visible droplets that we call "clouds." Clouds may vaporize again, or droplets may grow large enough to fall as rain, hail, or snow.

Cloud formation requires water vapor, but also aerosol particles for the vapor to condense around. These "seeds," or condensation nuclei, are extremely small (they might have to be enlarged 50 times to equal the size of a period on this page). Without aerosol particles, air could grow very humid, but the sky would remain clear. Even in "clean" desert or mountain air, there can be enough tiny dust, fungi spores, or transported salt particles to allow cloud formation.

The phrase "on cloud 9" originates from a classification system where "cloud type number 9" describes the tall, puffy cumulonimbus thunderheads. The simplest categorization begins with four types based upon shape, then adds terms that describe altitude and combines terms together to cover the range of characteristics (fig. 14).

"Cumulus" means "heaped" in Latin. Puffy cumulus clouds can grow over six miles high (up to 10 miles in the tropics). In

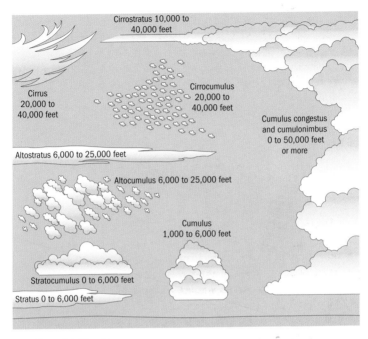

Cirrostratus 10,000 to 40,000 feet

Cirrus 20,000 to 40,000 feet

Cirrocumulus 20,000 to 40,000 feet

Cumulus congestus and cumulonimbus 0 to 50,000 feet or more

Altostratus 6,000 to 25,000 feet

Altocumulus 6,000 to 25,000 feet

Cumulus 1,000 to 6,000 feet

Stratocumulus 0 to 6,000 feet

Stratus 0 to 6,000 feet

Figure 14. Cloud types.

Plate 12. Cumulus clouds over the Mono Lake basin.

a sky with many cumulus clouds, generally all will have flat bottoms at the same elevation, where the air is cold enough for water vapor to condense and form clouds (pl. 12).

"Stratus" clouds are layered. Stratus often forms where warm air passes over colder air, causing moisture to condense at the cool boundary elevation (pl. 13). The gray "linoleum sky" of a Minnesota winter, as described by Garrison Keillor, is a stratus overcast. Sometimes such layers appear mottled where air rises or falls within the layer.

Plate 13. Stratus clouds on the Humboldt coast near McKinleyville.

"Cirrus" means "curled." Cirrus clouds form very high, five to eight miles up. They are ice crystals that appear ephemeral and wispy (pl. 14). The curls, elongated by wind, have been called "horsetails." Often they move very fast, traveling within high-altitude jet streams, but since they are up so high the movement appears slow from the ground. Cirrus clouds can signal the approach of storms because they form in advance of low-pressure fronts.

The height of the bases of clouds provides additional information. The prefix "alto-," as in "altostratus," is used for clouds at middle levels, from around 6,000 to 25,000 feet. Clouds whose names contain the prefix "cirro-" form at higher levels. There is no prefix for the lowest clouds.

With the basic terms, descriptive combinations become possible. For example, "altocumulus" are midlevel clouds that are lumpy (pl. 15), while "altostratus" are layered clouds at midlevel.

Plate 14. Cirrus clouds, with wispy "horsetails" stretched by the wind at high altitudes.

The added term "nimbus" describes clouds that produce rain. "Cumulonimbus" are thunderheads that may create rain, hail, and lightning (fig. 15, pl. 16). They may grow vertically to heights of more than 50,000 feet. Within the towering clouds are ascending and descending currents that create powerful turbulence that is dangerous to aircraft. Movement up, then down, and up again causes successive layers of freezing on ice droplets, until they become heavy enough to fall to

Plate 15. Altocumulus: puffy clouds at a midlevel altitude.

the ground as hail. In drier conditions, cumulonimbus clouds release "virga," or rain veils, in which rain begins to fall but evaporates before it reaches the ground.

Lightning, Then Thunder

In a typical thunderstorm, electrically charged particles build up in cumulonimbus clouds, possibly because of collisions during turbulence (the exact process remains mysterious).

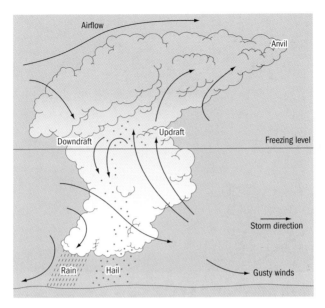

Figure 15. Thunderhead. Arrows show winds associated with a cumulonimbus thunderstorm.

Plate 16. Cumulonimbus thunderhead forming near June Mountain in the eastern Sierra Nevada.

Plate 17. A lightning down-strike at Mono Lake.

Plate 18. Cloud-to-cloud lightning over San Francisco.

Smaller particles tend to acquire positive charges and migrate toward the upper portions of the cloud, while negative charges on larger particles accumulate nearer the base. The cloud's negativity induces a concentration of positive charges along the ground below. When the charge difference grows great enough, electrons begin to descend out of the cloud, and positive charges are drawn upward, often channeled through vertical objects like trees, utility poles, or tall buildings. A visible lightning bolt occurs when the current jumps across the gap, flowing at 60,000 miles per second (about one-third the speed of light) in a series of upward and downward strokes (pl. 17). Our eyes cannot follow the rapid up and down movements, so we see bolts that flicker. Cloud-to-cloud lightning strikes are even more common than ground strikes (pl. 18).

In California, only 79 people died from lightning strikes in the three-plus decades between 1959 and 1994. During the same period, there were 1,523 deaths caused by lightning in Florida and 498 in Texas (the two leading states for lightning casualties). California experiences almost two strikes per square mile a year, which sounds like quite a lot, but most of these occur over mountainous parts of the state. In fact, the central California coast experiences the least lightning activity in the entire United States.

Thunderstorms in the Sierra Nevada typically happen late on summer days, when rising hot air and upslope winds lead to condensation and cloud formation. A monsoon circulation that draws moist air from the Gulf of Mexico into California during the summer leads to much of this thunderstorm activity in the mountains; sometimes (but less often) that unstable air can deliver lightning storms to other parts of the state.

Thunder is the sound produced by lightning. The heat generated by a lightning bolt makes the air expand explosively and then contract, generating sound waves. The sound travels about a mile in five seconds. Light travels much faster, so we see bolts before we hear thunder. Long rumbles occur when sound originates from different parts of the lightning bolt's path.

In its efforts to avoid injury or death by lightning, the National Weather Service warns against being the tallest thing in an open field or standing on mountain peaks or ridges during a thunderstorm. Taking shelter beneath isolated trees is also dangerous; instead, look for low areas or ravines, or move beneath thickets of small trees. Stay away from open water. If inside, keep away from windows and doors. Stay off the telephone and turn off computers and televisions; they can be damaged if utility lines are struck. It really is true that your hair may stand on end just before lightning is about to strike. That is an urgent warning to take shelter immediately or quickly crouch down with your feet together to reduce contact with the ground.

Clouds on Your Level

Fog forms when clouds are at ground level or just above it. As with all clouds, fogs take shape when water condenses around condensation nuclei. Coastal fogs take advantage of the abundant salt-spray aerosols near the shore, which facilitate condensation even when relative humidity is less than 100 percent.

San Francisco is famous for fogs manufactured by the coastal band of cold upwelling water and the moist air blown through the Golden Gate. The summer fog probably explains the droll assertion, often attributed to Mark Twain, that the coldest winter he ever spent was a summer in San Francisco. Twain never actually made that statement, but under-dressed summer tourists to the "City by the Bay" can appreciate the kernel of truth in that joke.

The early-summer fogs and overcast marine layer along the California coast are so predictable, and sometimes block the sun for so many days, that the phenomenon has been named "the June Gloom" (though it can persist into July and August too in the northern part of the state). An early travel

Plate 19. Tule fog in the Central Valley.

writer in southern California, J. Smeaton Chase, was fascinated by coastal fog in Laguna Canyon:

> Suddenly the sea-fog that lay continually in wait along the frontier of the coast, gaining a temporary advantage by some slackness of the enemy, poured over the mountain to the southwest and cast the whole mass into impressive gloom. On the instant the leaf was turned, brush was transmuted to heather, from California I was translated to Scotland. Fringes of sad gray cloud drooped along the summits or writhed entangled in the hollows of the hills. One who did not know the almost impossibility of rain at midsummer in this region would have declared that it was imminent. A strong breeze blew salty in our faces; but when by mid-afternoon we rode in the village of Laguna Beach, the sun held sway. So the unceasing warfare goes along this coast. (Chase 1913, 13)

The Central Valley experiences a winter wet-season fog called tule fog (pl. 19). In Eastern Sierra lake basins where snow

Plate 20. The freezing winter fog, "pogonip," typical of Great Basin valleys like the Mono Lake basin.

falls in the winter, a freezing fog forms during calm weather that decorates all objects with a rime of ice crystals. The native Paiutes called this fog "pogonip," a term that has been formally adopted by the National Weather Service (pl. 20).

Contrails

If someone from any time in history before the mid-twentieth century could see the California sky today, they would be amazed by the number of straight white lines of cloud criss-crossing the sky (pl. 21). Contrails form behind aircraft when warm exhaust mixes with cold air, causing water vapor to condense high in the sky. The process is analogous to seeing your breath on a cold day. These artificial clouds, manufactured by humans, are a very new phenomenon in the long history of the atmosphere.

Plate 21. A contrail, a manmade sky phenomenon.

Modern sky watchers can pay attention to the persistence of jet contrails for weather forecasting. On days where the vapor trails disappear quickly or do not form at all behind high-flying jets, dry weather is in store. If the air aloft is somewhat moist, contrails may form immediately behind the aircraft but last only a short while. In very moist air, contrails develop behind airplanes and persist in calm conditions, sometimes for many hours.

Long-lived contrails can grow wider and fuzzier as time passes. No longer straight white lines, they begin to look like normal cirrus clouds. At high altitudes, strong winds may move them far from where they first appeared. In 1996, NASA satellites tracked contrails over California and the southern United States that lasted as long as 17 hours and expanded to enormous sizes. One grew bigger than the state of Connecticut.

An apparent increase in the number of cirrus clouds during the last 30 years seems to correlate with increased air traffic. The effect is far greater in some places than in others. I live near Mono Lake, which is directly beneath the path of commercial flights east of the Bay Area. We watch contrails

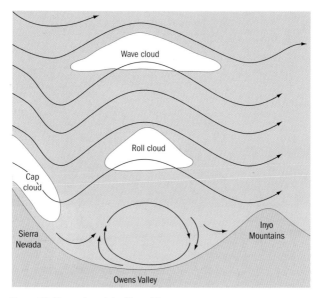

Figure 16. Formation of the Sierra Wave.

form by the dozens, particularly in the spring, and gradually spread across much of our sky.

When contrails lead to extensive cloud cover, do they cause cooling or warming of the ground below? That remains uncertain. White cirrus clouds reflect incoming sunlight (and prevent it from reaching the ground), but usually that cooling effect is more than canceled out when the "blanket" of cirrus traps surface heat and causes local warming. Contrails are denser, however, so their whiteness may reflect even more sunlight. An intriguing experiment occurred during the three days when all aircraft were temporarily grounded in the United States after the terrorist attacks of September 11, 2001. Without contrails to modify the weather, daytime temperature fluctuated higher, and nighttime temperatures were lower in areas that normally had lots of air traffic and contrail formation. The study, published in *Nature* in August 2002

Plate 22. The Sierra Wave forming lenticular clouds along the eastern crest of the mountain range.

(Travis et al. 2002), compared temperatures on those three days to average temperatures on those dates over many years, and to temperatures before and after the shutdown.

The Sierra Wave

Clouds may assume unusual shapes other than those in the classic categories. "Lenticular" clouds form when winds "stretch" clouds into lens, or saucer, shapes. The effect is not uncommon over the crest of the Sierra Nevada, where westerly winds hit the mountain range at right angles, are deflected upward, and then drop on the east side. The wind of the Sierra Wave keeps racing past, with a stationary lenticular cloud capping the crest of the standing wave (fig. 16, pl. 22). Often, a number of diminishing wave and roll clouds form. Their uplift helps gliders climb to extremely high altitudes (over 40,000 feet!) starting from the Owens Valley.

California's mountain ranges influence cloud formation

Plate 23. Air blown against mountain ranges rises and cools, and the water vapor it holds condenses to form clouds.

and precipitation because they generally run perpendicular to the prevailing westerly winds. The Coast Range, the Sierra Nevada, and southern California's Peninsular Ranges experience their heaviest precipitation on west-facing slopes, where air is forced upward. High elevations are less weighed down by the atmosphere than lower elevations, and therefore experience less pressure. The expansion of air, as surrounding pressure drops, causes cooling; colder air releases rain or snow. When air descends the eastern slopes, it becomes compressed and warms again, holding more moisture. The "rain shadows" created by that process explain the presence of deserts east of California's mountain ranges (pl. 23).

Air in Motion

Daily and Seasonal Winds

*The air runs a constant race with itself to move from
high to low pressure—the race of the wind.*

—BETH PRATT (2003)

Winds are the product of the sun's energy, the earth's rotation,
and the variety of landforms. They are labeled with the directions they come from: a north wind arrives out of the north,
for example.

The sea breeze is a west wind in California. During the day,
the land heats up more than the ocean. Warmer air rises and
is replaced by cooler air blowing off the ocean. A warm day
at the beach can suddenly turn cold in the afternoon when
the sea breeze kicks in. Stratus clouds may also accompany the
sea breeze inland. A land breeze reverses such flows, usually
at night, when the ocean is warmer than the land. However,
during California summers, the ocean is usually not much
warmer than the land, even at night. Occasional land breezes
are most likely to happen in fall and winter.

The "delta breeze," a particularly powerful sea breeze that
blows from San Francisco Bay, is funneled through the Carquinez Straits and drawn inland to displace hot air rising over
the Central Valley. The breeze usually arrives by evening and
brings cool relief to those it reaches. Such winds turn both
north and south, but seldom penetrate very far into the valley
in those directions.

Wildland firefighters learn to watch for changing wind
directions that affect fire behavior after the sun goes down.
During the heat of the day, valley-mountain winds blow up
the sides of heated slopes, often quite powerfully. In the
evening, upslope winds stop and reverse direction, usually
blowing more gently, because hilltops and slopes cool more
rapidly than valleys and ravines. Cooler air, then, sinks downward from the ridges.

Santa Ana and Diablo Winds

When high-pressure weather systems lie over the Great Basin and the Mojave deserts, air is pushed downward through mountain passes. That is when the powerful Santa Ana winds blow hot and dry, sometimes at more than 100 miles per hour, across southern California's coastal basins and out to sea. Descending air comes under pressure from the increased weight of the atmosphere overhead, so it gets compressed and heated (at a general rate of 5.4 degrees F per thousand vertical feet).

"Santa Anas" are a special category of land breeze, most noticeable in southern California in fall. It is no coincidence that the threat of wildfire peaks at the same time (pl. 24). Chaparral shrubs have adapted over the millennia to this regular pattern of wind-driven fire. The shrubs burn intensely hot, yet root crowns are designed to sprout after each fire, and

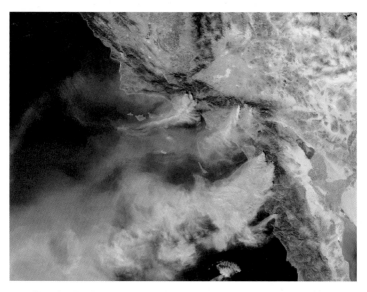

Plate 24. Ash from multiple wildfires in southern California, October 2003, blown by Santa Ana winds out to sea.

Plate 25. A dust devil along the shore of Mono Lake.

seeds of some plants require the heat of a fire to germinate. Ever since people began planting their houses and cities across California, we have struggled to learn how we too can become "fire-adapted" creatures.

While the winds blow, the moving air dominates everything and everyone in the region. During Santa Ana winds, novelist Raymond Chandler wrote, "Every booze party ends in a fight. Meek little wives feel the edge of the carving knife and study their husbands' necks. Anything can happen" (Chandler 1946, 11).

Occasionally, a similar wind pattern occurs in the San Francisco Bay Area, where it is called the Diablo Wind. High pressure inland pushes air out of the Central Valley from the direction of Mount Diablo, the dominant mountain peak east of the Bay.

Dust Devils and the Occasional Tornado

Tornadoes are violently rotating columns of air that extend, like gray funnels, from the bases of cumulonimbus clouds to the ground. They are relatively uncommon in California. The state averages only five per year and ranks forty-fourth among all the states in frequency per square mile. Occasionally, a powerful one appears in the state, but most are weak. The spinning of a tornado results from the updrafts and downdrafts in a thunderstorm interacting with different wind speeds and directions between the ground and aloft. Now and then, pilots report seeing tornadoes above the ground, where they are called funnel clouds.

Dust devils are much more common in this state. They are not tornadoes. Instead of descending from clouds overhead, they form on bright, sunny days. A layer of heated air develops near the ground, rises, and begins spinning due to the prevailing winds. Dust devils often appear over the plowed fields of the San Joaquin Valley before crops emerge and on the white salt flats of the shoreline around Mono Lake (pl. 25).

Blowing in the Wind

Wind is a force of nature that sculpts the landscape and even shapes ocean waves. Surfers know that offshore winds help form waves into the ideal shape for their sport. Wind blowing on the coast and in the deserts moves sand and piles it into dunes (pl. 26).

Plate 26. Eureka sand dunes, sculpted by wind in Death Valley National Park.

Plate 27. Monterey cypress sculpted by prevailing coastal winds at Point Lobos.

Places characterized by regular strong winds reveal that fact in their vegetation. Plants that grow on alpine ridges or along windy beaches may hug the ground closely to avoid the perpetual assault or, when they venture some vertical growth, take on shapes that bow to the power of air in motion. Botanists call the resulting tree postures "krummholtz," a German term meaning "crooked wood," or "cushion plants" for plants growing above treeline. Writer J. Smeaton Chase described the result at Cypress Point; there, Monterey cypress grow on headlands "where winds careen most wildly [and] the gaunt wardens of the cliff have been torn, twisted, hunched, wrenched, battered, and hammered to the limit of tree resemblance" (Chase 1913, 175) (pl. 27).

Pollination, Planting, and Perfumes

Wind is also a force that plants and animals have turned to their advantage in order to survive and reproduce. Through the air, plants and animals exchange chemical communications that allow them to influence the behavior of others and to advertise their presence to prospective mates. Pheromones are chemicals that send long-distance signals to trigger programmed responses. They have been called "the smell of love." Male moths may track a pheromone trail upwind for miles to reach the female sending out her message. Farmers control pests like the codling moth, the proverbial worm in the apple, by using pheromone traps in orchards to disrupt mating of the pests. This targets the specific problem in a way that is far more benign than pesticide applications.

Social order is maintained by many animals using "social odor." Whether chemical signals come from specialized scent glands or are secreted as part of sweat or urine, these communications define territorial boundaries, signal social status, or advertise availability and readiness to breed. Most people are well aware of how much time and effort dogs devote to scent-marking and to studying such messages from other dogs. This is a habit common to all canines, including coyotes and wolves. If we could detect the invisible messages ourselves, perhaps it would be easier for dog owners to understand why their pets are not allowed on trails in most state and national parks, where the local wildlife takes precedence.

People also send chemical signals, for similar reasons. The marketing of perfume is one obvious example of how we rely on scent to communicate. We most likely send subtle messages via pheromones too (those that we do not overwhelm with applications of other scents, that is).

Air disperses perfumes and transmits colorful images from flowers, effectively announcing: "nectar available here." Those who respond find a sweet food source as they get dusted by pollen and fertilize the flowers in return. Most pollinators are insects that respond by flying toward the flowers' signals. The air, then, becomes the medium both for marketing and delivery in this round-about mode of sexual reproduction.

The other grand strategy for pollination is to take direct

advantage of the wind itself. The flowers of wind-pollinated plants are located near the tips of branches or stems to best catch the breeze. Such plants scatter enormous numbers of pollen grains into the air, as every hay fever sufferer knows. They usually have separate male and female flowers, often on different individual plants. The pollen grains themselves are tiny, to facilitate long-distance voyages, and come with adaptations for flight like wings or fluff. At the other end of the trip, flowers capture the male pollen with relatively large stigmas—the female landing targets—and optimize exposure to the rain of pollen by having small petals or none at all. Wind pollinators are not in the market for insect visits, so they do not usually bother with flashy, colorful displays or nectar.

Dry or temperate climates suit this approach better than wet zones do, since rain washes pollen from the air before it can travel very far. So it is no surprise that wind pollination evolved as a common tactic for plants across most of the dry state of California. Many of the dominant plant types in California plant communities are wind pollinators. Male catkins produce the pollen on coast live oak trees *(Quercus agrifolia)*. Female flowers appear later and are less noticeable. Pines (*Pinus* spp.) and other conifers send out yellow clouds of pollen early each summer. Some of this pollen turns the edges of lakes yellow when it washes into quiet eddies or splashes onto beaches. Willows (*Salix* spp.), cottonwoods (*Populus* spp.), cattails (*Typha* spp.), many desert shrubs, and all grasses are wind pollinated (pl. 28).

After successful pollination, plants face the challenge of seed dispersal. Once again, the wind is harnessed for this purpose. Some seeds have extensions that resemble parachutes. Dandelion seeds, for example, sport umbrella-like clusters of hairs to carry them on the wind. Other plants send out "helicopter seeds." If you pull the scales from a ponderosa pine *(Pinus ponderosa)* cone, for example, you will find a nut surrounded by an oblong wing that rotates as it falls through the sky, lofting the seed out from under the shade of its parent

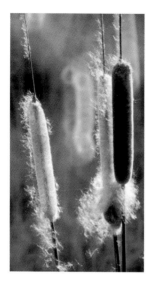

Plate 28. *Left:* Cattails releasing seeds to the wind. *Below:* A dandelion, ready for a breeze to carry its seeds aloft.

tree. After a forest fire, this adaptation allows the wind to carry new seeds into the burned zone.

Trees in the elm family produce seeds surrounded by papery wings that send them spinning through the air. Desert shrubs like four-winged saltbush *(Atriplex canescens)* use the same kind of design. One cattail spike may contain a million tiny seeds. Each seed has silky white hairs to catch the wind. The fluffy seeds have been used for waterproof insulation and to make life jackets buoyant. Cottony seeds are also characteristic of willows, California sycamores *(Platanus racemosa),* and cottonwoods, the trees that annually release white "cotton" that may blanket the ground like early-summer snow.

The desert tumbleweed *(Salsola tragus)* takes another approach, using the wind not for flight but propulsion along the ground, and scatters seeds as it goes. Tumbleweeds are more accurately called Russian thistle. Though they have become ubiquitous symbols of our drylands, the species is not native to California, but arrived with the Spanish explorers.

Motion in Air: Taking Flight

Designing California Birds

To overcome gravity and take flight, birds have light, hollow bones, airfoil-shaped feathers and wings, and aerodynamic bodies. Birds do not paddle the air with their wings, as if pushing at water from a rowboat. Instead, they cut the air with the leading edge of wings to make it flow over them. All airfoils work because of their curvature, which forces air traveling above a wing to move farther and faster than air passing underneath (fig. 17). With less pressure above than below, a bird (or an airplane) can rise through the air. Birds maneuver and adapt to constantly changing air conditions by adjusting their wing shapes and feather positions with a sensitivity that airplane pilots can only envy.

Three birds found in California help illustrate the range of

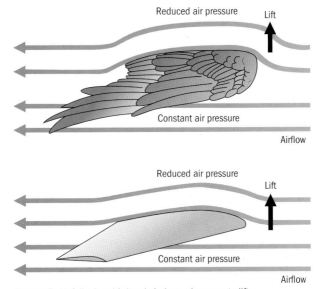

Figure 17. Air foils: how bird and airplane wings create lift.

adaptations that birds have made to a life in the sky. Strikingly different wing designs are tied to different strategies these birds use to find food and travel through the sky.

California Condors

One spring day, as I was preparing this book, we watched several Turkey Vultures *(Cathartes aura)* in wobbly flight over a windy ridge near Big Sur. They held their wings in the characteristic V-shape as they battled a stiff breeze. Suddenly, three

Plate 29. A soaring California Condor.

other birds came cruising toward us from the coast. Their long, broad wings were extended flat from their bodies; they flew straight and stable as Stratocruisers, despite the gusts. The three huge California Condors *(Gymnogyps californianus)* passed directly overhead. After condors had been nearly lost to extinction, a captive breeding program became successful enough to release a breeding population of the birds in the Big Sur wilderness.

Condors soar on warm thermal updrafts for hours (pl. 29). When they decide to travel long distances, they might climb to altitudes of 15,000 feet. Speed is not particularly important to them as they cruise, watching for carrion, but winds may carry them along at up to 55 miles per hour. They have been known to fly 150 miles in one day.

Like all vultures, condors more typically loiter in the air, gaining lift from masses or columns of rising warm air. They have the broad wings characteristic of soaring birds, but theirs are huge: a nine-foot wing span! Such big birds ought to be ungainly, but that adjective only applies on the ground. In the air, they are agile, thanks to "slotting" along the wing tips, where individual feathers are adjusted for maneuverability in the air.

To get off the ground at all, condors often must wait until the sun has heated the ground enough to generate thermals. Winds pushing against and over hills generate lift that also lets the birds do "slope soaring." They sometimes hover in one place, riding such stationary waves.

California Gulls

For most gulls, it is not flying that matters, but eating. For this gull, though, it was not eating that mattered, but flight. More than anything else, Jonathan Livingston Seagull loved to fly.

— RICHARD BACH (1970, 12)

Plate 30. California Gulls have wings suited for speed, maneuverability, and distance.

Built for high speed, the wing of a California Gull *(Larus californicus)* (pl. 30) shows a much slimmer profile, for its length, than a condor wing. Gull wings are swept back at an angle, instead of perpendicular to the bird's body (as is the case with most condors). They have sleek edges without slotting, a design suited for fast flight and aerial prowess. Gulls are omnivores that chase down prey like live fish, but also maneuver for carrion or trash in competitive flocks, when stealing a neighbor's catch may require adroit speed. They take advantage of thermals too, over the ocean or the land. We often see them inland in circling flocks near Mono Lake, gaining thousands of feet in altitude before breaking out of the circle to cruise toward the high-altitude meadows and lakes of the Sierra Nevada.

California Quails

California Quail *(Callipepla californica)* require very different capacities for flight than vultures and gulls do. They have elliptical wings, short and wide, with many slots that help them get off the ground and maneuver in short bursts of fast flight (pl. 31). The official "state bird" has been clocked flying 58 miles per hour, but quail are sprinters rather than long-distance flyers. Though quail can deftly maneuver around and between the branches of trees and shrubs, they usually only fly as a last resort, preferring to run away along the ground.

Plate 31. A California Quail's rapid flight.

Can Bumblebees Fly?

The old myth that bumblebees should not be able to fly shows how little we understood about the physics of flight. Insects were the first creatures to evolve designs for taking to the air. The bumpy-looking bee wing or a butterfly's seemingly simple wings are actually efficient airfoils, able to constantly adapt to the complexities of moving air. With such wings, Monarch Butterflies *(Danaus plexippus)* make 3,000-mile migrations. Dragonflies, such as the California Darner *(Aeshna californica),* are flying acrobats whose wings simultaneously twist while beating, so they can hover or even fly backward. Pairs of dragonfly wings work independently, with the front ones rising as the rear ones fall.

To overcome gravity, insects benefit from being such lightweights. However, that also creates flight problems. The lightest wind will force mosquitoes out of the air, bringing relief to their victims but leaving the insects hungry. Inevitably, when the wind dies down, they quickly appear in flight again, following trails of exhaled carbon dioxide upwind toward their food sources.

Spiders do not fly, but in fall, young ones sometimes disperse to new homes on gossamer webs. They spin web parachutes, hold them out where the wind can sweep them aloft, then sail through the air for great distances. Like adaptations for seed dispersal in plants, the spiders use wind to spread from their places of birth and extend their species' range.

Mammals Overhead

Bats are the only mammals in the world capable of true flight. Their bodies are flying sails: leathery wings feature a membrane stretched across delicate finger and arm bones. Bats wiggle their fingers to change direction while chasing flying insects. Being mammals, bats have fur instead of feathers, but their wings are naked.

The 24 species of bats in California play an essential role in

Plate 32. California Leaf-nosed Bat homing in on a cricket.

controlling insect numbers. They navigate and find food with echolocation, high-pitched sound waves they send out through the air to bounce off objects. Bats are not blind, but actually see and hear very well.

The California Leaf-nosed Bat *(Macrotus californicus)* has better night vision than military night scopes (pl. 32). It sneaks up on flying moths and then, at the last second, uses echolocation. These bats make their homes in desert habitats from Riverside, Imperial, San Diego, and San Bernardino counties south to the Mexican border.

Humans have done their best to join bats, their fellow mammals, and the other flyers in the sky. John Balzar eloquently described the feeling of piloting a small airplane: "There is a certain grace, mechanized but still grace, that comes with sitting at the controls . . . of a small airplane . . . carving gentle arcs through the troposphere, riding an un-

Left: Plate 33. Woman with a kite at the San Ramon Wind and Art Festival.
Right: Plate 34. Balloons setting sail at the San Ramon Wind and Art Festival.

seen cushion of nitrogen and oxygen. Your machine, so thin-skinned and fragile on the ground, conforms nobly to its gaseous surroundings" (2004).

Our fascination with flight led us to invent airplanes, balloons, kites, hang gliders, and parasails and to celebrate the air, the wind, and flight in annual festivals and shows (pls. 33, 34). Thousands of people gather at such events, where professional kite flyers may control three at once, making them dance to music; where colorful balloons are inflated at dawn and sail away, trusting to the wind; and where screaming jets and burbling World War I aircraft cavort overhead or are on display on the ground.

During World War II, a half-million pilots were trained by the military, many at California air bases. After the war, the aerospace industry found its home in southern California.

Individual piloting of planes never did reach the peaks predicted during the 1950s, when some envisioned helicopters on the roof of every garage and daily aerial commutes for most of the populace. Commercial air travel has grown tremendously, of course, but the experience in such aircraft seems designed to separate us as much as possible from any sensations of flight. Even the windows are, more often than not, curtained to facilitate in-flight movies.

Yet the fascination with flight remains. Private pilot licenses, issued by the Federal Aviation Administration, are available after 40 hours of flight instruction and supervised practice. Recreational pilot certificates allow pilots to carry passengers during daytime hours within 50 miles of their home airport after only about 30 hours of training.

Southern California is considered America's capital of aviation. That fact is partly explained by the character of the air in the southern California air basins.

The purity of the air in Los Angeles is remarkable.
The air . . . gives a stimulus and vital force which only
an atmosphere so pure can ever communicate.

—BENJAMIN C. TRUMAN (MAJOR) (1874, 33–34)

A TOUR OF CALIFORNIA can provide an overview of the ways geographical features shape air movement while exploring the climate variations and diverse character of air around the state (map 3). Areas that share the same air because of topography, weather, and climate conditions are called airsheds. The State of California has been officially divided into fifteen "air basins" (map 4). The basins begin with the airshed concept, then factor in political boundaries to create manageable bureaucracies. The state's air basins are further divided into either county Air Pollution Control Districts (APCDs) or regional Air Quality Management Districts (AQMDs).

Such man-made boundaries for air basins and air districts understate the actual movements of air. Passes and river canyons channel air through mountain barriers from one basin into another. Air masses can also be lofted high overhead so that winds can carry persistent air pollutants hundreds or even thousands of miles from their origins. All of California's major urban areas contribute to air pollution in other air basins and, at times, receive transported pollution.

Our tour begins on the coast, starting from the south, where so many of the state's air quality issues originate. The route then moves inland, to basins in the Central Valley that receive air and pollution transported from the coast, but where locally generated pollution adds to the mix, which is then passed along to mountain basins. We will explore the air basins along the northeastern and eastern edges of the state, where marine influences come to an end and Mediterranean climates give way to continental patterns. Finally, to the south are California's desert air basins, downwind from the southern coast.

San Diego County

Ask residents of San Diego County why they live there, and almost certainly they will mention the region's "perfect" weather and climate. The city of San Diego and its suburbs form the third-largest urbanized area in California after the

Map 3. Landform provinces in California.

South Coast and the San Francisco Bay Area (pl. 35). San Diego is home to industrial and transportation facilities, military installations, an international airport, and a major shipping port.

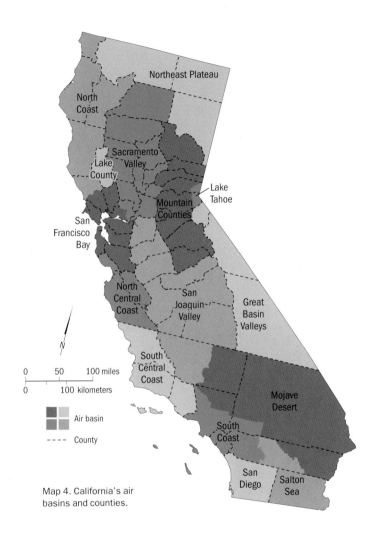

Map 4. California's air basins and counties.

Legend:
Air basin
- - - County

Prevailing daytime winds blow from the ocean eastward across a coastal plain to the peninsular ranges that form the interior boundary of the basin. To the south lies the international border with Mexico. San Diego shares air, and emis-

Plate 35. The urbanized coastal plain of San Diego.

sions, with the city of Tijuana, immediately across the border. Air pollution may also be transported down from the South Coast, although local emissions alone can cause violations of air quality standards.

South Coast

The South Coast Air Basin (map 5) experiences the weather and climate that epitomizes California in the imaginations of most people around the world, who have seen southern California's sunny skies and striking beaches in Hollywood movies or tourist brochures. Since the late nineteenth century, a concerted effort has been made to market the region as a destination for tourists and immigrants. Agricultural opportunities brought early settlers. Once the citrus industry developed, colorful crate labels advertised the oranges and lemons they contained and also evoked visions of the region as an idyllic escape from snowy winters into "perpetual" sunshine. An 1873 book promoting the region included a letter from Anaheim that extolled the region's air:

Southern California presents a most gloriously invigorating, tonic, and stimulating climate, very much superior to any thing I know of, the air is so pure and much drier than at Mentone [on the Italian Riviera], . . . it has a most soothing influence on the mucous membrane, even more so than the climate of Florida, and without the enervating effort of that. (Nordhoff 1873, 248–249)

Settlers were attracted to the region because there was more truth than hyperbole in such praise. The mild climate remains attractive. Most of the coastal valleys have freeze-free seasons that last 225 to 300 days per year. Los Angeles sees more days (73 percent) of sunshine each year than any other major urban area in the nation except Phoenix. However, with more than 15 million people crowded into the South Coast region today, "invigorating" and "pure" air is no longer a selling point for southern California.

This fully urbanized coastal plain is where air quality control began, as California responded to the first smog episodes during World War II and more challenges during the latter half of the twentieth century. The expertise and commitment

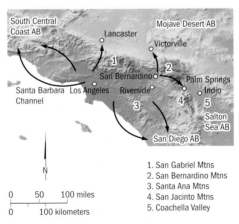

Map 5. South Coast Air Basin and transport patterns.

Plate 36. A smoggy sunset in the Los Angeles basin.

of the South Coast Air Quality Management District (SCAQMD) became a model for the nation. If the air could be cleaned up here (the thinking went), it might be cleaned up anywhere.

The SCAQMD has jurisdiction over the South Coast Air Basin, plus the Riverside County portions of the Salton Sea Air Basin and the Mojave Desert Air Basin—a total of 10,743 square miles. The basin includes parts of four counties (all of Orange County and the nondesert portions of Los Angeles, Riverside, and San Bernardino counties).

The Los Angeles region helped put the word "inversion" into the public's vocabulary. Inversions and the marine layer are dominant weather features much of the year, especially from late spring through early fall (pl. 36). Warm summer air commonly lies over the cooler marine layer, trapping pollution. Typically, along the South Coast, the layer of cool, humid marine air advances inland at night and in the early morning, then "burns off" to hazy sunshine midday.

The air basin is bounded by the Pacific Ocean on the west,

by the San Gabriel and San Bernardino mountains (the Transverse Ranges) to the north and east, and by the San Jacinto and Santa Ana mountains to the east and south. Sea breezes and the prevailing westerly winds push air from Los Angeles inland to concentrate pollution against the mountains in places like Pasadena and Pomona, or farther east around Riverside and San Bernardino. Some of the earliest evidence that ozone pollution damaged pine forests began to appear in both the Angeles and San Bernardino national forests.

Air can then escape through several gaps in the barrier, moving in three different directions. It may be channeled south along the inland side of the Santa Ana Mountains toward Escondido and the foothills east of San Diego. A forest of wind turbines marks another path eastward through San Gorgonio Pass, past Palm Springs and Indio, followed by a southward turn into the Coachella Valley and the Colorado Desert. The third route starts north of San Bernardino, at the intersection of the San Gabriel and San Bernardino mountains, following the Interstate 15 freeway over Cajon Pass and

Plate 37. A clear winter day near Devore, where the interstate approaches Cajon Pass.

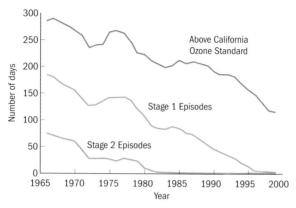

Figure 18. Ozone trends in the South Coast Air Basin, 1965–2000.

into the eastern Mojave Desert around Victorville and Hesperia (pl. 37).

North of Los Angeles, Soledad Pass sends air and pollution through a gap in the San Gabriel Mountains out to the western Mojave Desert. Air can also blow along an inland route from the San Fernando Valley up to Ventura and Santa Barbara counties. An offshore route develops during Santa Ana wind episodes, which blow out to sea and may then turn north or south. When the typical onshore breeze resumes, it then delivers the South Coast's pollution to Santa Barbara or Ventura counties or south to San Diego County.

The South Coast Air Basin's issues illustrate, better than anywhere else, the connection between population numbers and air quality concerns (fig. 18). Since the 1970s, when ozone levels peaked, the number of days with Stage 1 or 2 smog alerts has declined significantly. Yet,

> increases in the population over that time have made overall emission reductions more difficult. . . . Unless significant steps are taken to further control air pollution, growth will overwhelm much of the improvements expected from the existing control program. (SCAQMD 2003, 1–4)

South Central Coast

The South Central Coast Air Basin includes San Luis Obispo, Santa Barbara, and Ventura counties. The coastal plain is narrower here than in the Los Angeles/Orange County region to the south. Rugged ridges isolate and limit airflow between interior valleys.

Santa Barbara and Ventura counties contain power plants, oil extraction and oil refining facilities, and transportation and agricultural activities (pl. 38). Although there are no major commercial ports, large vessels passing through the Santa Barbara Channel produce roughly one-third of the nitrogen oxide emissions in Santa Barbara County, more than all such emissions from cars, trucks, and buses combined. Santa Barbara and Ventura are the only counties in the state with high risk from radon, the radioactive gas emitted by certain soils that may build up indoors and present a risk for lung cancer.

Plate 38. Santa Barbara, where onshore winds generate cumulus clouds over the coastal hills.

Plate 39. Morro Bay in San Luis Obispo County. Note the power plant stacks on the right.

San Luis Obispo County is separated from the other two counties by the Santa Ynez Mountains. Much of the population lives in the city of San Luis Obispo and nearby coastal communities (pl. 39). Paso Robles, farther north, is isolated from the coast by mountains, and its air patterns are more closely linked with the San Francisco Bay Area and San Joaquin Valley.

North Central Coast

Monterey Bay is the dominant coastal feature in the North Central Coast Air Basin, which consists of Santa Cruz, San Benito, and Monterey counties. From the mouth of the Salinas River, the Salinas Valley extends southeastward, spanning 25 miles at the coast and gradually narrowing to less than four miles wide at King City, 60 miles inland (pl. 40). South of Monterey Bay, the Santa Lucia Mountains define the western boundary of the Salinas Valley. To the north, the Santa Cruz

Plate 40. The Salinas River valley.

Mountains separate the region from the San Francisco Bay Area. A gap in the Diablo Range opens at Pacheco Pass, where turbines generate electricity as wind blows out to the San Joaquin Valley.

Sea breezes and coastal fogs follow the same basic patterns as elsewhere on the coast. Robert Louis Stevenson wrote eloquently about the summer fog moving inland at sunset:

> From the hill-top above Monterey . . . the upper air is still bright with sunlight; a glow still rests on the Gabelano Peak; but the fogs are in possession of the lower levels. . . . It takes but a little while till the invasion is complete. The sea, in its lighter order, has submerged the earth. Monterey is curtained in for the night in thick, wet, salt, and frigid clouds. . . . And yet often when the fog is thickest and most chill, a few steps out of the town and up the slope, the night will be dry and warm and full of inland perfume. (Stevenson 1990 [1895], 162)

The air basin's largest population centers are Santa Cruz, Salinas, and Monterey. The basin has relatively clean air.

Plate 41. Steam rises from the Moss Landing power plant on Monterey Bay, which is fueled by natural gas.

Emissions problems are generally localized near pollution sources like a cement plant, the Moss Landing power plant, agricultural activities, and highway traffic corridors (pl. 41).

Scotts Valley, along the Highway 17 corridor in the Santa Cruz Mountains, can receive transported air and pollution from the San Jose area. If wind currents lift air higher aloft, out of the Bay Area, transported air can impact Pinnacles National Monument in the mountains south of Hollister.

San Francisco Bay

The San Francisco Bay Air Basin occupies a central location on California's coast (map 6). It is the second-largest urban area in the state after the South Coast. The basin includes Alameda, Contra Costa, Marin, Napa, San Francisco, San Mateo, Santa Clara, and Solano counties, as well as southern Sonoma County.

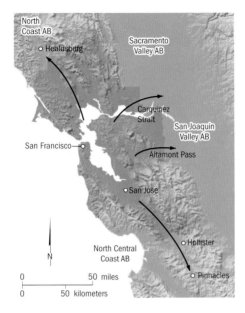

Map 6. San Francisco Bay Area Air Basin and transport patterns.

Westerly sea breezes help maintain good air quality for most of the Bay Area (pl. 42). On typical summer mornings, the entire city of San Francisco lies in the fog. Later in the day, skies on the inland side of the bay usually clear up as winds blow in from the Pacific Ocean, but only partial clearing occurs on the cooler ocean side. Strong sea breezes, which commonly blow 20 to 30 miles per hour in the afternoon, channel through the Golden Gate and then out the mouth of the Sacramento River at the Carquinez Strait or over Altamont Pass, east of Livermore. At Altamont Pass, electricity-generating wind turbines are positioned along ridges to take advantage of the predictable winds.

In 1868, John Muir praised the air in the South Bay region that today encompasses the solidly urbanized Silicon Valley:

> The sky was cloudless, and the whole [Santa Clara] valley was a lake of light. . . . The San Jose sky was not simply pure and

bright, and mixed with plenty of well-tempered sunshine, but it possessed a positive flavor, a *taste* that thrilled throughout every tissue of the body. Both my companion and myself had lived on common air for nearly thirty years, and never before this discovered that . . . this mortal flesh, so little valued by philosophers and teachers, was possessed of so vast a capacity for happiness. (Badè 1923, 178–179)

Plate 42. An unusually smoggy day in San Francisco.

Pockets of redwood forest correspond with the coastal fog belt north and south of the bay. Portions of this air basin are also famous for microclimates that make the Napa and Sonoma Valleys suitable for wine growing (pl. 43). The valleys are close enough to the bay for fog to penetrate, but far enough inland to provide a warm growing season with cool nights. Under such conditions, grapes ripen slowly to maximize flavor. The balance between cooling fog and warm sunshine influences growers' choices of grape and wine varieties. Black or red grapes are favored over white grapes in areas with warmer climates. The summer drought is also a key to the success of the region's wineries; rain at the wrong times is a critical factor in wine production.

In winter, fog forms in the moist regions of the Sacramento–San Joaquin River delta and moves out to San Fran-

Plate 43. Napa Valley vineyards depend on a balance between sunny, hot days and cool fog drifting in from San Francisco Bay.

cisco Bay on cool, easterly breezes. This type of fog is less common than summer's variety, but it can be very dense and cause traffic problems on the freeway system. It can also carry particulate pollution from the Sacramento Valley into the Bay Area.

Spring and fall bring the hottest days and the most sunshine, as high pressure builds into the Pacific Northwest and the Great Basin. Under those conditions, dry offshore winds may replace the Pacific sea breeze. Prevailing summer winds can blow from the urban areas of Santa Rosa and Petaluma, north of San Francisco, toward Healdsburg, in the North Coast Air Basin.

North Coast

The wettest, most unvarying climate in California can be found along the North Coast. Parts of this air basin receive 60 to 100 inches of rain per year. Sometimes rain is falling only in this part of the state while the rest of California has fair weather. Weather here is strongly influenced by the ocean and

Plate 44. The Mattole River valley in Humboldt County.

the rugged topography of the mountain ranges. The coast has cool, foggy summer days and mild winters.

The air basin extends down the coast from the Oregon border, varying from 30 to 100 miles in width. It includes northern Sonoma County and Mendocino, Humboldt, Trinity, and Del Norte counties. The basin's biggest cities are Ukiah, Willits, Eureka, and Crescent City.

Here the Coast Range and Klamath Mountains merge in rugged forest terrain, cut by deep river canyons that extend inland more than 200 miles (pl. 44). Winter snows can accumulate on the higher peaks, where elevations exceed 9,000 feet in the Klamath Mountains.

Locations of coast redwood *(Sequoia sempervirens)* forests correspond with the fog belt (pl. 45). Though most of the precipitation falls in winter, as is typical for the rest of the state, summer fog-drip is an important part of the redwood trees' seasonal water intake.

Known as a city with the coldest summers in the nation,

Plate 45. Coast redwoods are the dominant trees in California's rainforests.

Eureka receives only 41 percent of the possible sunshine during winter because of clouds or fog, and its annual average is just 51 percent. But across the region, this air basin enjoys some of the best air quality in the state, aided by winds off the ocean.

Lake County

The only air basin in California that meets all air quality standards is the Lake County Air Basin (pl. 46). The basin includes just one county, Lake County, 80 miles north of San Francisco. The dominant geographic feature of the county is Clear Lake and its surrounding watershed. There are no large urban zones upwind of Clear Lake, and the lake basin has a relatively small population. But vehicles driven in the basin do generate nitrogen oxides that may have contributed to elevated levels

Plate 46. The clean skies of Clear Lake.

of nitrogen deposited into the lake. Occasionally, particulate smoke from forest fires blows in from the North Coast. A wind-energy farm may someday occupy the ridge tops straddling the line between Lake and Colusa counties.

Sacramento Valley

The Sacramento Valley is more than 150 miles long and averages about 50 miles wide (pl. 47). The valley is ringed by the Coast Ranges to the west, the Cascades to the north, and the Sierra Nevada to the east. These surrounding mountain ranges form natural barriers to air movement, although river canyons and mountain passes channel winds. There are no significant barriers to north-south air movement within the entire Central Valley (map 7).

With ideal soils and climate for crops, the valley is one of the leading agricultural regions in the nation. This northern half of the Central Valley gathers water in the Sacramento River from a number of tributaries; the water flows out

Plate 47. Looking north up the Sacramento Valley.

through the Sacramento–San Joaquin delta toward San Francisco Bay. The Sacramento River's riparian zone and marshlands, now mostly converted to irrigated farmlands, help humidify the air on the valley floor. Humidity is highest during winter and lower during the heat of summer.

The delta breeze blows strongest on summer evenings, delivering cooler air that penetrates toward the city of Sacramento, then continues northward up the valley or east to the foothills of the northern Sierra Nevada. It can also carry pollution from the San Francisco Bay Area. The Sacramento Municipal Utility District maintains wind turbines near Rio Vista, in Solano County, to take advantage of the strong spring and summer delta breezes, when Sacramento most needs added power.

On some summer days, winds sweep south out of the

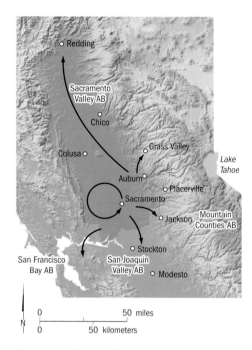

Map 7. Sacramento Valley Air Basin and transport patterns.

Sacramento Valley into the northern San Joaquin Valley. Afternoon breezes from the west may then push polluted air from the valley up into the Sierra Nevada foothills.

Another characteristic wind pattern of the southern Sacramento Valley is a counterclockwise eddy that circulates northwest from Sacramento and turns south to Woodland and Davis, then back east to Sacramento.

Winters are wet and cool. After winter storms, calm weather may bring tule fogs that persist for long periods until another storm event arrives. These valley fogs usually stay below the 1,000-foot elevation, so that residents of foothill towns like Auburn, on sunny winter days, may look out over the top of a cloud layer filling the bowl of the valley.

The air basin takes in parts of 11 counties: all of Butte, Colusa, Glenn, Sacramento, Shasta, Sutter, Tehama, Yolo, and

Yuba counties, plus the western part of Placer and the eastern part of Solano counties. Because air masses often behave differently in the north and south sections of the valley, the basin has been broken into the Broader Sacramento Area and the Upper Sacramento Valley planning areas. Air pollution transported from the Broader Sacramento Area dominates the air quality in the Upper Sacramento Valley as far north as Butte and Tehama counties. However, air quality in Shasta County, at the far northern end of the Sacramento Valley, may be the product of local emissions or entirely due to transport from the south, or sometimes a mixture of both.

San Joaquin Valley

The San Joaquin Valley stretches for 300 miles, about one-third the length of the state, and ends at the San Gabriel and Tehachapi mountains in the south (map 8, pl. 48). Three population clusters are spread widely apart, north to south: Stockton/Modesto, Fresno/Visalia, and Bakersfield. Much of

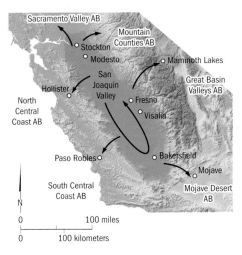

Map 8. San Joaquin Valley Air Basin and transport patterns.

Plate 48. A hazy day in the southern San Joaquin Valley.

the air entering the valley first travels through the Bay Area and across the delta toward the urban areas of Stockton and Modesto. With rapidly growing populations, all of the valley's major cities add their own impacts to the region's air quality.

The air basin includes Fresno, Kings, Madera, Merced, San Joaquin, Stanislaus, and Tulare counties, plus the western half of Kern County. Most of the valley's industry is related to agriculture, although forest products and oil production/refining add to local emissions.

Pacheco Pass sends air from the Hollister area into the San Joaquin Valley; Pacific Gas and Electric operates wind turbines in that pass. Winds leaving the southern end of the valley through Tehachapi Pass blow into the Mojave Desert (turning turbines owned by Southern California Edison).

Winds also blow eastward up the canyons of the Sierra Nevada during the day, as far as the crest of the Sierra. In summer, air currents transporting pollution into the foothills of Sequoia National Park can make the air quality there as bad as any we find in the worst parts of the South Coast Air Basin.

Cool winds descending into the valley from the mountains can create a circular pattern on the valley floor, known as

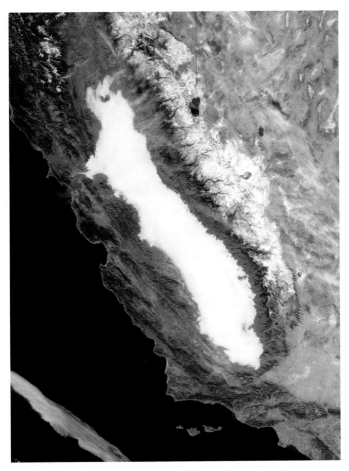

Plate 49. Tule fog fills the San Joaquin Valley and the southern Sacramento Valley, January 17, 2003.

the Fresno eddy. The eddy swirls pollution along the east side of the valley from Tulare County and northern Kern County back to the Fresno area.

For anyone driving north into the San Joaquin Valley from

southern California, over the "Grapevine" Pass on Interstate 5, it is not uncommon in the winter to look over the sunlit tops of a solid mass of ground-hugging tule fog and then descend into suddenly reduced visibility (pl. 49).

In December 2003, San Joaquin Valley air authorities made the historic request to be downgraded into the "extreme" polluter category, voluntarily joining Los Angeles as the country's worst offenders of federal ozone standards. Mobile sources were responsible for most of the standards violations and were proving difficult to clean up, because the local district had no authority over engine and fuel standards. District officials were looking forward to 2007, when more stringent federal and state requirements on vehicles and fuels would take effect. Their request pushed the cleanup deadline back from 2005 to 2010, avoiding annual fines that would have otherwise been levied in those intervening years. While the San Joaquin Valley has actually had more "bad air" days in recent years than the Los Angeles region, its peak levels of smog have been lower than the peaks in southern California.

Mountain Counties

The Mountain Counties Air Basin covers the central and northern parts of the Sierra Nevada, from the foothills to over 10,000 feet at the crest of the Sierra. The air basin includes Amador, Calaveras, Mariposa, Nevada, Plumas, Sierra, and Tuolumne counties, plus the middle portions of El Dorado and Placer counties. Throughout the region, economies that once were based on logging and mining in Sierra Nevada communities have refocused on tourism and recreational activities. Increasing numbers of people living in the air basin commute to jobs in the Central Valley.

Snow falls in moderate amounts nearly every winter at elevations as low as 2,000 feet. Above 4,000 feet, snow can remain on the ground for most of winter. Afternoon and evening

Plate 50. Near the Placer County Courthouse in Auburn, Interstate 80 carries heavy volumes of vehicle traffic through the Sierra Nevada foothills.

thunderstorms occur in summer, when rising heat causes cumulonimbus clouds to form over the peaks. Lightning strikes often start wildfires. Smoke from seasonal fires is a natural part of the mountain air regime, but after a century of fire suppression the character of fires and smoke episodes has changed. Accumulated fuels have led to a relatively few massive firestorms that produce enough smoke to be seen from space and to spread sometimes across the western states. These regional smoke episodes are more extreme than the smoke from low-intensity ground fires that used to be common in Sierra Nevada forests.

Vehicles speeding across the air basin on Highway 50 or Interstate 80, or winding through foothill communities connected by Highway 49, account for significant portions of local pollutant emissions (pl. 50). However, most of the basin's air pollution comes from the Sacramento Valley, the Bay Area, the San Joaquin Valley, or combinations of all three upwind air basins.

Yosemite National Park, in parts of Mariposa and Tuolumne counties, experiences less severe air quality issues than parks in the southern Sierra, yet its problems are also significant (pl. 51). Protecting the special purity of mountain air is a challenge for the National Park Service, and it becomes ever more difficult for modern visitors to appreciate what Bolton Coit Brown felt in 1896:

> Such vast gulfs of air above, below, and around; such wealth of warm sunshine; such a paradise of sunny solitude sweeping aloft far into the sky's deepest blue—these and the intoxication of the air at ten thousand feet . . . stirred in me a deeper sense of the *heavenliness* of the mountains and a deeper joy in them than was ever mine before. (Brown 1896, 298)

Plate 51. High-country vista in the Tuolumne River watershed.

Lake Tahoe

Lake Tahoe sits within a true basin, 6,000 feet above sea level, surrounded by peaks of the Sierra crest. The air basin includes portions of El Dorado and Placer counties, while the eastern

Plate 52. Lake Tahoe, where development contributes to air pollution and reduces clarity in the lake.

half of the basin lies in Nevada. Pine forests cover the slopes of the "bowl." More than 20 million visitors come to Lake Tahoe each year to participate in winter sports, to go camping and fishing in summer, and to gamble in casinos on the Nevada side of the lake. Most visitors drive cars up to the lake and around its shores. They use fireplaces in rental cabins, condominiums, or vacation homes.

Tahoe is renowned for its incredible water clarity, but automobile emissions and wood burning have increased measurable levels of nitrogen, ozone, acid rain, and other pollutants (pl. 52). Over half the nitrogen entering Lake Tahoe comes from the air itself. Unnatural fertilization by such deposits contributes to eutrophication, an increase in algae that diminishes the lake's clarity. Polluted air also blows into the basin from outside. Toxic organic compounds have shown up in lake trout, and sediments sampled from the bottom of the deep lake show elevated concentrations of mercury and lead.

Northeast Plateau

With a climate regime that is distinct from the rest of California, the Northeast Plateau Air Basin includes Siskiyou, Modoc, and Lassen counties (pl. 53). The basin has sharply defined seasons that follow a continental, rather than marine, pattern. Winters are cold and snowy, summers warm and dry.

Plate 53. Looking east, toward Susanville, a part of California that has a continental, rather than a Mediterranean, climate.

The basin includes part of the Klamath Mountains to the west and the Cascade Range and Modoc Plateau, plus a slice of the Great Basin, along its eastern edge. Mount Shasta rises 14,162 feet, dominating the view from much of the basin. Another volcanic peak, Mount Lassen, stands 10,457 feet high. Extensive forest land runs across saddles between the region's peaks. The volcanic Modoc Plateau extends across the northeastern expanse, with an elevation mostly above 4,500 feet.

The region receives no transported air pollution from major urban areas (pl. 54). As in many rural areas in California, particulates from dust and wood smoke are sometimes a problem. Only the city of Yreka experiences occasional ozone concentrations that approach "near exceedances."

Plate 54. A rainbow near Tule Lake in northeastern California's typically clean air.

Great Basin Valleys

Elna Bakker, whose book explores the variety of California's habitat types, has characterized the Great Basin habitat on the eastern border of California as "a harsh climate, but very seasonal—a wonderful region for homesick easterners to visit when they complain of the climate placidity of coastal California" (Bakker 1971, 236). All life in the Great Basin, as Bakker put it, has had to adapt to a "lack of climatic coddling."

This is a land of contrasts. Along the great wall of the Sierra Nevada, Pacific storms drop winter snow on mountain peaks, but a few miles to the east, precipitation drops significantly and the vegetation turns to Great Basin sagebrush scrub. The air basin encompasses Alpine, Inyo, and Mono counties (pl. 55). It includes Death Valley and the lowest point in the United States, 282 feet below sea level. The same air basin takes in Mount Whitney—at 14,494 feet, the highest peak in the 48 states. Mono Lake sits at 6,400 feet and experiences a climate that prompted Mark Twain, in 1872, to joke: "There are only

Plate 55. Long Valley and Crowley Lake in the clear skies of the Eastern Sierra.

two seasons in the region round about Mono Lake — and these are, the breaking up of one winter and the beginning of the next." (Twain 1973 [1872], 246). (That year was near the peak of a global cooling episode that came to be known as the Little Ice Age.)

South of the Mono Basin, the dominant feature of the air basin is the long, narrow Owens Valley. The "rain shadow" effect of the Sierra Nevada means that the southern parts of the Owens Valley average only four inches of rainfall per year. The Owens River runs the length of the valley. However, water diversions to Los Angeles that began early in the twentieth century dried both the river and Owens Lake, at the south end of the valley.

The bed of Owens Lake extends about 17 miles north and south and 10 miles east and west, covering approximately 70,000 acres. Once it was exposed, winds began lofting dust and salts from the lakebed. Fine-particulate episodes in the Owens Valley violate air quality standards, and at times, levels have been the highest ever measured in the nation. The air

Plate 56. Dust blowing off the dry bed of Owens Lake.

standard can be exceeded more than 50 miles away, with dust affecting residents from Ridgecrest to Bishop. Many visitors come to the Eastern Sierra and high desert for their recreation attractions and end up spending time in the impacted area. The City of Los Angeles Department of Water and Power has been required to install dust control measures on Owens Lake (pl. 56).

Most of the rest of the air basin enjoys good air quality, although winds over the mountain passes sometimes bring pollution from the Fresno area into Mammoth Lakes. That resort town is the largest in the air basin and generates its own local emissions primarily due to wood burning in the winter. Mammoth Mountain is a volcanic feature with unusual carbon dioxide issues: on some parts of the mountain, forest trees have died due to carbon dioxide gas emerging from the soil. Signs warn visitors not to linger in those localities, where exposure to the gas could also be deadly to humans (pl. 57).

Plate 57. A carbon dioxide warning on the south slope of Mammoth Mountain; note the dead tree, one of many killed by CO_2 released from the local soil.

Mojave Desert

The Mojave Desert Air Basin occupies the California "high desert," which lies mostly between 2,000 and 4,000 feet above sea level. Joshua trees *(Yucca brevifolia)* and creosote bushes are the predominant vegetation. As in all deserts, the Mojave's dry air heats rapidly during the day, then cools rapidly at night. Winds, sand dunes, and blowing dust are typical of deserts, with their sparse vegetation. Lots of sunshine is also common, making deserts natural places to site solar electric-generating plants (pl. 58).

The basin includes the eastern half of Kern County, the northern part of Los Angeles County, most of San Bernardino County except for the southwest corner, and the eastern edge of Riverside County. It is separated from the San Joaquin Valley, to the west, by the Tehachapi Mountains and the southern end of the Sierra Nevada.

Plate 58. Photovoltaic solar panels at Kramer Junction in the Mojave Desert.

Most of the basin is sparsely populated, but the area just north of the San Gabriel and San Bernardino mountains has a large, fast-growing population made possible by water delivered from northern California via the California Aqueduct (pl. 59). Military bases, highways and railroad facilities, cement manufacturing, and mineral processing add local emissions to the basin's air.

Plate 59. Palmdale, growing in the Mohave Desert with water from the California Aqueduct.

Salton Sea

In the southeast corner of the state, most of the Salton Sea Air Basin takes in California's "low desert," below 1,000 feet in elevation and marked by the Sonoran-type desert vegetation that extends into Arizona and Mexico. Temperatures are hotter here than in the Mojave Desert. Winds blow primarily from the south and west. The mouth of the Coachella Valley, at San Gorgonio Pass, is one of the major outlets for air from the South Coast (pl. 60).

Plate 60. Wind turbines generating electricity near Palm Springs.

Irrigation water from the Colorado River and a year-round growing season that allows farmers to harvest multiple crops support a multibillion-dollar farm economy. Most of the population resides in the Coachella Valley and in the Imperial Valley, south of the Salton Sea. There are no large cities in the region, but modest-sized communities include Palm Springs, Indio, El Centro, and Calexico. Cities in the Coachella Valley are experiencing some of the most rapid population growth rates in the state. The city of Mexicali lies just across the border from Calexico. With a population of about

Plate 61. The Salton Sea, a migratory stop in the desert for millions of birds.

one million, it sometimes impacts air quality on the California side of the border.

The basin is in "nonattainment" status for violations of particulate standards. In this region, high childhood asthma and death rates from respiratory diseases have at times been twice the statewide rates, according to the California Department of Health Services. Wind-blown dust particulates are suspected as the major culprit of these health issues. That problem could be aggravated by moving water out of the Imperial Valley to San Diego. Should the Salton Sea drop 15 feet, the water loss would expose about 70 square miles of lakebed (pl. 61). Dust blowing off that salt flat could increase exposure to unhealthy particles small enough to be drawn deep into the lungs. Ironically, the coastal cities' thirst for population growth would then lead to more air quality problems for people in the distant agricultural valley.

We have met the enemy and he is us.

—POGO (KELLY 1970)

CALIFORNIANS HAVE CHANGED the air, leaving a human "footprint" in the atmosphere that affects everything and everyone living in the state. Although Californians can take pride in the progress made fighting air pollution and in leading the nation to face such challenges, the majority of Californians still breathe air with unhealthy levels of pollutants. Some of the state's air basins have the dubious distinction of "leading the nation" when it comes to dirty air. Studies suggest that breathing air in parts of southern California can reduce one's life expectancy by more than two years.

The Breath of Life

We take a breath about every five seconds. Every day, an adult breathes in and out over 3,000 gallons of air. Children breathe even more air in proportion to their smaller body weights. They breathe faster, particularly during strenuous physical activity. They also spend more time outdoors than any other segment of the population and tend to breathe through their mouths, bypassing the natural filtering of air pollutants by the nose. For similar reasons, athletes may be among the most vulnerable to the effects of air pollution, even though they are relatively young, healthy, physically fit, and nonsmokers.

A 10-year health study that monitored 5,500 southern California children was completed in 2004. It was the longest study ever done on the impact of air pollution on developing lungs. The study analyzed effects of particulates, ozone, nitrogen dioxide, and acid vapor in 12 southern California communities (some that were heavily polluted and others with cleaner air). The children received annual health examinations, and their school absences were documented. Housing characteristics (private residences versus apartment complexes, for example) and exposure to tobacco smoke in homes were factored into the analysis.

This long-term study found that active children living

in communities with the dirtiest air, including Long Beach, Mira Loma, Riverside, San Dimas, and Upland, were up to three times more likely to develop asthma. By age 18, 8 percent of children who grew up in those communities developed less than normal lung capacity (below 80 percent of normal capacity). Living near freeways with high traffic levels led to higher risks. School absences due to respiratory illnesses like runny noses and asthma attacks went up after air pollution events. Children who moved to cleaner communities during the study showed improved lung function.

What's a ppb to Me?

Federal and state standards have been set for traditional, or "criteria," pollutants, such as ozone, particulates, carbon monoxide, oxides of nitrogen, sulfur dioxide, and lead. The term "criteria" comes from the fact that the U.S. Environmental Protection Agency (EPA), under the law, must describe the criteria—the characteristics and potential health and welfare effects—applicable to each pollutant. Standards also define the number of exceedances permitted before a violation occurs. California's own standards are usually more stringent than the federal ones, and the state has additional standards for vinyl chloride, hydrogen sulfide, sulfates, and visibility-reducing particles for which no federal standards have been set. Other air pollutants have been classified as "toxic air contaminants" (TACs) that may cause serious, long-term health effects even at low levels. Both criteria pollutants and TACs are measured statewide (table 3).

The relative number of pollution molecules in an air sample is measured in parts per million (ppm) or parts per billion (ppb). Particulate matter, however, is measured as micrograms found in each cubic meter of air (mg/m^3).

In 1999, the EPA developed the Air Quality Index (AQI) to simplify the daily reporting of air pollution levels to the

TABLE 3. State and Federal Air Quality Standards

Pollutant	Averaging Time	California Standard[*]	Federal Standard[*†]
Ozone (O_3)	1-hour	.09 ppm (180 µg/m^3)	.12 ppm (235 µg/m^3)
	8-hour	——	.08 ppm (157 µg/m^3)
Respirable Particulate Matter (PM_{10})	24-hour	50 µg/m^3	150 µg/m^3
	Annual mean	20 µg/m^3	50 µg/m^3
Fine Particulate Matter ($PM_{2.5}$)	24-hour	No separate state standard	65 µg/m^3
	Annual mean	12 µg/m^3	15 µg/m^3
Carbon Monoxide (CO)	8-hour	9.0 ppm (10mg/m^3)	9 ppm (10 mg/m^3)
	1-hour	20 ppm (23 mg/m^3)	35 ppm (40 mg/m^3)
	8-hour (Lake Tahoe)	6 ppm (7 mg/m^3)	——
Nitrogen Dioxide (NO_2)	Annual mean	——	.053 ppm (100 mg/m^3)
	1-hour	.25 ppm (470 mg/m^3)	——
Sulfur Dioxide (SO_2)[†]	Annual mean	——	.030 ppm (80 mg/m^3)
	24-hour	.04 ppm (105 mg/m^3)	.14 ppm (365 mg/m^3)
	1-hour	.25 ppm (655 mg/m^3)	——
Lead	30-day average	1.5 mg/m^3	——
	Calendar quarter	——	1.5 mg/m^3
Visibility-reducing Particles	8-hour	In sufficient amount to reduce the prevailing visibility to less than 10 miles when relative humidity is less than 70 percent.	——
Sulfates	24-hour	25 mg/m^3	
Hydrogen Sulfide	1-hour	0.03 ppm (42 mg/m^3)	——
Vinyl Chloride	24-hour	0.01 ppm (26 mg/m^3)	——

[*] Concentrations are expressed first in the units in which they were promulgated. Equivalent units in parentheses are based upon a reference temperature of 25 degrees C and a reference pressure of 760 torr.

[†] National primary standards are shown; they are levels of air quality necessary, with an adequate margin of safety, to protect the public health. The 3-hour averaging time for sulfur dioxide has no primary standard, but a secondary one of .5 ppm (1,300 mg/m^3). National secondary standards are levels necessary to protect the public welfare from any known or anticipated adverse effects of a pollutant.

public. The AQI combines levels of ozone, particulates, carbon monoxide, sulfur dioxide, and nitrogen dioxide in a single numerical, color index (table 4). The higher the index number, the higher the pollution level and greater the likelihood of health effects. Specific cautions are provided for people in "sensitive groups" (those with lung or heart diseases, and children or adults who are physically active outdoors).

The AQI replaced an earlier Pollution Standards Index, which is still calculated daily for metropolitan areas with populations over 200,000. Under that system, when pollution reaches very unhealthy levels, Stage I smog alerts are issued to

TABLE 4. Air Quality Index

Index Values	Air Quality Description (Color)	Health Cautionary Statement
0-50	Good (Green)	No limitations.
51-100	Moderate (Light Yellow)	Extremely sensitive children and adults, especially those with respiratory diseases such as asthma, should consider limiting outdoor exertion.
101-150	Unhealthy for Sensitive Groups (Orange)	Sensitive children and adults, especially those with respiratory diseases such as asthma, should limit prolonged outdoor exertion.
151-200	Unhealthy (Red)	Sensitive children and adults should avoid outdoor exertion, and everyone else should limit prolonged outdoor exertion during peak ozone periods.
201-300	Very Unhealthy (Purple)	Sensitive children and adults should avoid outdoor activities and remain indoors. Everyone else should avoid outdoor exertion.
Over 300	Hazardous (Deep Purple)	Everyone, especially children, should avoid outdoor activities and remain indoors.

warn everyone to avoid strenuous outdoor activities; under Stage II smog alerts, everyone is advised to remain indoors as much as possible.

The Pollutants

Photochemical Smog: NO$_x$, VOCs, and Ozone

Photochemical smog is the air pollution type first identified in Los Angeles and southern California (pl. 62). The series of smog reactions is complex, but overall can be summarized as:

$$\text{VOCs} + \text{NO}_x + \text{heat} + \text{UV sunlight} \xrightarrow{\text{O}_2} \cdots \rightarrow$$
$$\text{ozone (O}_3\text{)} + \text{nitrogen dioxide (NO}_2\text{)} + \text{hydrocarbons}$$

Volatile organic compounds (VOCs) are carbon molecules that are sometimes referred to as "reactive organic

Plate 62. Los Angeles, lost in smog and particulate haze; the view from Mt. Wilson in August 2004.

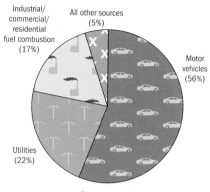

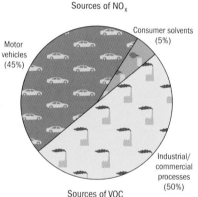

Figure 19. Sources of NOₓ and VOCs.

gases" or "reactive hydrocarbons." They are produced in vehicle exhaust and industrial emissions. Nitrogen oxides (NO_x) are compounds such as nitric oxide (NO) and nitrogen dioxide (NO_2), typically created during combustion processes. Thousands of tons of both NO_x and VOCs are emitted into California's atmosphere every day (fig. 19).

VOCs drive the smog reaction by supplying the oxygen atom that converts NO into NO_2. The NO_2 molecules then enter ozone-forming reactions when acted upon by heat and sunlight. California's famed sunshine and warm weather are its curse when it comes to smog formation. Below 90 degrees F, smog is seldom a problem. Above 95 degrees F, the South Coast Air Basin suffers from unhealthy air nearly every day.

A series of reactions forms ozone, a strong-smelling, pale blue gas that is reactive and toxic. Ultraviolet rays in sunlight first separate an oxygen atom (O) from NO_2 to form nitric oxide (NO) and a free oxygen atom. When the free oxygen atom encounters one of the air's abundant oxygen molecules, ozone (O_3) forms.

Other by-products include stable hydrocarbons. Some of these are toxic air pollutants, and others are aerosol particles that form haze. One is the gas PAN (peroxyacetylnitrate), which irritates eyes and lungs.

Nitrogen dioxide absorbs blue light, so we see it as the brownish-red color of smoggy air, as if the sky itself were developing an unhealthy suntan. Nitrogen dioxide combined with water vapor in the air becomes nitric acid, which also stings eyes and throats. Acid can fall directly onto the ground or be carried by rain back to the land, where it can turn surface waters acidic.

Ozone's Health Effects

The American Lung Association (ALA) annually lists the top 25 metropolitan areas in the United States with the worst ozone pollution. In 2004, seven of the top eight were in California: (1) Los Angeles–Riverside–Orange County, (2) Fresno, (3) Bakersfield, (4) Visalia-Porterville, (6) Sacramento, (7) Merced, and (8) Hanford-Corcoran. Two other California areas had the dubious honor of making this list: (18) San Diego–Carlsbad and (20) Modesto.

When listed by county, California's status looked even grimmer. Thirteen of the nation's 25 worst ozone polluting counties were in this state, including 11 of the top 13: (1) San Bernardino, (2) Fresno, (3) Kern, (4) Riverside, (5) Tulare, (6) Los Angeles, (8) Merced, (9) El Dorado, (10) Kings, (12) Nevada, (12) Sacramento, (17) Ventura, and (23) Placer.

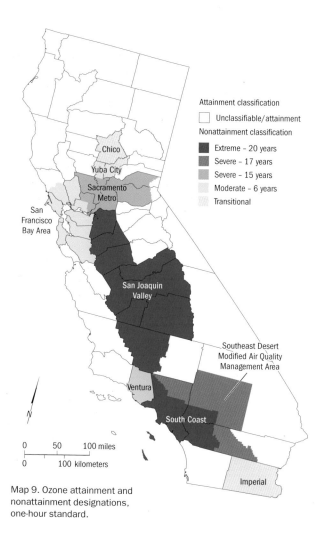

Attainment classification

☐ Unclassifiable/attainment

Nonattainment classification

■ Extreme – 20 years

■ Severe – 17 years

■ Severe – 15 years

■ Moderate – 6 years

▨ Transitional

Map 9. Ozone attainment and nonattainment designations, one-hour standard.

The ALA is concerned about ozone because, at ground level, it causes a number of adverse health effects. Being exposed to ozone concentrations as low as 90 ppb to 120 ppb for as little as one hour can impair lung function (map 9). Ozone injures the alveoli and bronchioles in the lungs and aggravates diseases like asthma, bronchitis, and emphysema. Elevated ozone levels have been correlated with increases in hospital admission rates and increased mortality rates. Studies document athletic endurance beginning to suffer once the ozone level reaches .12 ppm (the federal health standard), then dropping off more significantly at .20 ppm for athletes exercising heavily for one hour or more.

Children represent only 20 percent of the resident population in the South Coast Air Basin, yet they experience more than 40 percent of the symptoms associated with exposure to ozone. The most active children who spent the most time outdoors showed the greatest ill effects.

Ozone is also harmful to plants and wild animals. With elevated ozone levels, pine needles lose chlorophyll and forest trees die or are weakened enough that bark beetles, drought, and fire can combine to cause extensive mortality (pl. 63).

Plate 63. Dead trees in the San Bernardino National Forest, near Lake Arrowhead; weakened by smog and drought, they lose their ability to fight off bark beetles.

Agricultural crop losses due to ozone were first documented in southern California in 1949; now they cost an average $700 million per year.

The Mysterious "Weekend Effect"

The relationship between NO_x and VOCs in ozone formation is complex. Nitrogen dioxide is first a product in the smog reactions, as VOCs react with NO. It is also consumed, however, in the photochemical reactions that form ozone. When VOCs are abundant, NO_x promotes ozone formation, but when there are fewer of the volatile organic gases around, nitrogen oxides actually *restrict* the formation of ozone.

This complicated relationship may explain the perplexing "weekend effect." In recent years, ozone levels, though much lower overall than they were several decades ago, have been higher on Saturdays and Sundays than on weekdays in the state's major urban areas. This was unexpected, since commuters, big trucks, and industrial polluters are less active on weekend days. Over a period of five years in Los Angeles County, 260 days exceeded federal ozone standards, and 43.5 percent of these fell on weekends. In the South Coast Air Basin, ozone levels have sometimes been 32 percent higher on weekends than on Fridays. In the San Francisco Bay Area, levels have been 25 percent higher. To a lesser degree, the effect has also appeared in Sacramento and in the San Joaquin Valley.

Ozone differences on the weekend are puzzling. The explanation may have to do with the fact that, on weekends, the level of NO_x in the air has fallen more than the level of VOCs. Altering the ratio of the two compounds may cause *more* ozone to form. Some researchers think that we might be able to control ozone more effectively if we focus less on reducing nitrogen oxides and more on adjusting VOCs.

Experts disagree on this subject, however. The California Air Resources Board does not feel there is enough certainty yet to justify reducing the effort to control nitrogen oxides.

Ozone levels have declined on weekends, after all, just not as much as on weekdays. This has become a regulatory controversy, with auto industry lobbyists telling state legislators, during hearings in 1998, that SUVs (theoretically used more as weekend vehicles) should not have to meet the same NO_x emission standards as "everyday" cars.

Particulates

Particulates are very small aerosols that pose a health hazard, in part because of their chemistry but primarily due to their small size. They are categorized by size as either "respirable particulates," PM_{10} (less than 10 microns in diameter), or "fine particulates," $PM_{2.5}$ (smaller than 2.5 microns). Ten microns is only one-seventh the diameter of a human hair. Both types are tiny enough to penetrate deep into the lungs, but even more of the ultrafine particles find their way past our defenses. Our upper respiratory tract can trap most particles using mucus and cilia (small hairs that line our airway passages). We cough most of this material back out. Small particles make it past those defenses and move deep into the lungs, where they accumulate. Breathing them reduces lung capacity by coating the surfaces of bronchioles and alveoli, the smallest air passages in the lungs, where gas is exchanged with the blood. This lost lung capacity may be permanent.

Exposure to particulates can increase the number and severity of asthma and bronchitis attacks. According to recent research, death rates have increased 2 to 8 percent for every 50 mg/m^3 jump in exposure, even during short-term pollution episodes (note that 65 mg/m^3 is the federal standard for 24 hours of exposure for $PM_{2.5}$). Particulates cause breathing difficulties in people with heart or lung disease. Those with such conditions, along with children and the elderly, are most sensitive, but even healthy people are affected during high-concentration events.

Particulates enter the air from road dust, when farm and construction equipment stir dust into the air, and in smoke

Plate 64. Billowing smoke from a fire at Cuyamaca, east of San Diego, in 2003.

and ash from residential fireplaces, agricultural burning, or wildfires (pl. 64). Incomplete combustion of any burning fuel sends particulates into the air, so our cars, trucks, and factories are once again primary sources. Particulate aerosols also form when other pollutants in the air undergo chemical reactions. For example, when NO_x changes from a gas to nitric acid, it may then combine with sea salt or ammonia to form small particles.

The ALA ranks the metropolitan areas and counties in the United States that are most polluted by particulates (map 10). In the category "short-term particle pollution" (exposure to $PM_{2.5}$ over 24 hours) in 2004, 10 of the top 25 metropolitan areas were in California, including the worst three of all: (1) Los Angeles–Long Beach–Riverside, (2) Fresno-Madera, and (3) Bakersfield. Others that made the list from this state were: (8) Sacramento-Arden-Arcade-Truckee, (9) Visalia-Porterville, (11) Modesto, (12) Hanford-Corcoran, (15) San

Map 10. Particulate matter (PM$_{10}$) attainment and non-attainment designations.

Inside the map:

Attainment classification

☐ Unclassifiable/attainment

Nonattainment classification

■ Serious

▨ Moderate

Sacramento County

Mono Basin

Mammoth Lake

San Joaquin Valley

Owens Valley

Coso Junction

Trona

San Bernardino County (part)

South Coast

Coachella Valley

Imperial Valley

N

| 0 | 50 | 100 miles |

| 0 | 100 kilometers |

Jose–San Francisco–Oakland, (16) San Diego–Carlsbad–San Marcos, and (23) Merced.

The ALA also ranked metropolitan areas polluted by *year-round* particle pollution, and seven areas from California made that list, with the top four from this state. Happily, one city in California, Salinas, did make the list of the top 25 *cleanest* cities for year-round particle pollution.

As with the ALA's ranking for ozone, the list of counties most polluted by annual particle pollution was dominated by California, with the worst six counties being Riverside, San Bernardino, Los Angeles, Tulare, Kern, and Fresno. Also making the list were (8) Orange, (12) Kings, (20) Stanislaus, and (22) Merced.

Particles that occur naturally in the air and also affect humans include pollen, spores, salt from ocean spray, blowing dust, volcanic ash, and smoke from wildfires.

Carbon Monoxide

Carbon monoxide comes right out of our cars' tailpipes, so it is a "direct" pollutant (unlike photochemical smog, made indirectly through a series of reactions). It is a colorless, odorless gas produced by incomplete combustion of fossil fuels. Over 80 percent of the carbon monoxide emitted in urban areas comes from motor vehicles. Carbon monoxide binds to the blood's red blood cells so that they can no longer carry oxygen. Everyone learns not to run a car for very long in a garage or to cook with a barbecue indoors, because carbon monoxide can build up to deadly levels in confined spaces, but many don't realize that carbon monoxide can reach unhealthy concentrations outdoors too. Research has correlated carbon monoxide exposure in southern California to premature births, babies born with low birth weights, and cardiovascular birth defects. As of 2003, the South Coast Air Basin was one of the few areas in the country still designated as "nonattainment" for failing to achieve national standards for carbon monoxide.

Sulfur Dioxide

Sulfur dioxide (SO_2) is a strong-smelling, colorless gas formed when fossil fuels that contain sulfur are burned. Power plants that use coal or oil with a high sulfur content are the major sources of SO_2. (Fortunately, most of the fossil fuel power plants in California burn natural gas, rather than coal or oil.) An irritant gas that is usually filtered by our nasal passages, sulfur dioxide can become more of a threat when exercise, such as a brisk walk, raises the respiratory rate a moderate amount and triggers mouth breathing. When we breathe through our mouths, we can pull more sulfur dioxide down into the lungs. Even with brief exposure, SO_2 causes constriction of the airways that, in turn, results in wheezing, chest tightness, and shortness of breath. Such symptoms can appear at low levels of exposure for those with asthma, but can affect everyone at very high pollution levels. This pollutant is a particulate that also contributes to problems of "acid rain," when acid from the air is deposited onto land and water.

Coal-burning power plants generate sulfur dioxide. While California relies less on coal than other states and meets the federal standards for this pollutant, some of *our* sulfur dioxide has become a problem across a number of southwestern states. The Southern California Edison Company and the Los Angeles Department of Water & Power own 56 percent and 10 percent, respectively, of the Mohave Generating Station. This power plant in Laughlin, Nevada (near the point where the borders of California, Nevada, and Arizona intersect), has at times produced 3 percent of southern California's electricity and generated power for Nevada and Arizona. While making electricity, it also has created 40,000 tons of sulfur dioxide each year. The plant, fueled by coal slurry, has been the primary source of a white haze that, in the past, has spread across northern Arizona and southern Utah and obscured views of the Grand Canyon (pl. 65).

A lawsuit filed in the late 1990s by the Grand Canyon Trust and the Sierra Club was resolved by a consent decree that required installation of pollution control equipment by 2006. The owners of the plant pledged to spend $1.1 billion for new pollution control improvements that would add sulfur

Plate 65. The coal-powered Mojave Generating Station, sending soot into the air. The power plant ceased operations in 2005.

dioxide "scrubbers" and remove 85 percent of the sulfur from smokestack emissions, that would create a "baghouse" to capture 99 percent of the remaining particulates, and that would install burners to reduce nitrogen oxide emissions. However, Southern California Edison, the majority owner, closed the plant at the end of 2005 when it failed to meet the deadline for retrofitting.

Lead

Health effects from exposure to lead include kidney damage and a whole host of neurological impairments that can lead to seizures, mental retardation, learning disabilities, or behavioral disorders. Even low-level exposure to lead can damage the nervous systems of fetuses and young children.

In 1992, California finished phasing out lead as an ingredient in gasoline for on-road vehicles. The elimination of leaded gasoline reduced lead levels in the air by over 98 percent. But other sources still expose us to lead, including paints, tires, and fuels for off-road vehicles like aircraft, boats, and construction and farm equipment. Currently, fuel for private, propeller-driven airplanes is the primary source of inorganic lead. Leaded paints were banned in residences back in 1978 but are still allowed in industrial, military, and marine settings. Lead is listed both as a criteria air pollutant and as a toxic air contaminant.

Toxics

Toxic air contaminants (TACs) cause or contribute to an increase in deaths or serious illness, or pose hazards to human health. Health effects due to TACs may occur at extremely low levels. In fact, the safe levels for most TACs are not known. TACs are regulated under a different process than other air pollutants, within the California Health and Safety Code. California's current TAC list includes 243 substances. In California, one in every 15,000 people in the state—66 cases per million residents—is at risk of developing cancer from breathing airborne chemicals. The risk is much higher in some parts of the state—as high as 93 per million in Los Angeles County, according to an EPA analysis of 1999 pollution levels (table 5).

A few examples from the long list include hexavalent chromium, used in chrome plating and anodizing shops; perchloroethylene, used by thousands of dry cleaners in the state; dioxin, which shows up in medical waste incinerators; and trichloroethylene (TCE), used for metal degreasing, in dry cleaning, and for printing inks and paints.

Polychlorinated biphenyls (PCBs) can be found in industrial and consumer products, such as cooling compounds, many electronic instruments, paints, varnishes, plastics, inks, and pesticides. Some PCBs imitate reproductive hormones; even in very small concentrations, they can change wildlife reproductive capacity, longevity, intelligence, and behavior, or can lead to cancer or mutations.

TABLE 5. Airborne Chemicals Posing a Cancer Risk

Chemical	Main source	Cancer cases per million people
Benzene	Vehicles	14.43
Butadiene	Vehicles	6.41
Ethylene dibromide	Pesticides	6.11
Tetrachloroethylene	Dry cleaners, aerospace	5.20
Acetaldehyde	Vehicles	4.29
Tetrachloroethane	Various industries	4.01
Napthalene	Vehicles	3.95
Hydrazine	Rocket fuel	3.40
Carbon tetrachloride	Various industries	3.18
Hexavalent chromium	Metal finishing plants	2.34
Total risk across California		66.32

These are 10 of 171 chemicals evaluated at 1999 levels by the EPA.

Benzene is at the top of the EPA's list of cancer-causing TACs. It is a common solvent found in gasoline, inks, paints, plastics, and rubber. It also shows up in detergents, pharmaceuticals, and dyes. Exposure to benzene irritates the skin and eyes. Inhalation at high levels causes dizziness, weakness, headache, nausea, blurred vision, respiratory distress, and liver and kidney damage. Benzene has also been tied to diseases of the blood system and bone marrow. Controlling benzene requires regulations for over 14,000 service stations in California and for cars and light trucks.

Formaldehyde is, unfortunately, very common in our indoor air. It arrives via foam insulation, walls, and furniture made of particle board, plywood, or pressed wood. Paper products as seemingly innocuous as grocery bags, waxed papers, facial tissues, and paper towels can emit formaldehyde fumes. It appears in household cleansers, enables permanent-press fabrics to resist wrinkling or repel water, shows up in fire retardants, and explains much of that "new carpet smell." High formaldehyde levels in enclosed spaces irritate eyes, noses, and throats, in addition to causing headaches and aggravating asthma; they may also be a culprit in throat cancer.

Mercury is a persistent toxin that accumulates in the food chain. It enters the air from municipal and medical waste incinerators, coal-fired power plants, and other industrial sources. Breathing mercury in the air itself is usually not directly dangerous. However, mercury is delivered to the ground by rain and snow and enters lakes, streams, and estuaries. In the water, it changes into a very toxic form, methylmercury, which builds up in fish and animal tissues. People are exposed to mercury primarily by eating fish. Developing fetuses are most sensitive to the toxic effects. Children exposed to low concentrations while in the womb suffer an increased risk of poor attention spans and declines in fine motor function, language skills, visual-spatial abilities, and verbal memory.

According to the EPA, we have cut our emissions of mercury in the United States by more than 90 percent. Coal-fired power plants remain the largest domestic sources of mercury emissions (again, we do not use much coal in California plants). But the EPA estimates that about half of the mercury deposited in the United States comes from outside the country. It can be transported thousands of miles in the atmosphere, so effective control will require reductions in global emissions.

Radon is a colorless, odorless gas formed by the natural radioactive decay of thorium and uranium in soil or rocks. The gas evaporates out of the soil and can accumulate inside buildings. Exposure to the radioactive gas causes thousands of preventable lung cancer deaths. Radon is the second leading cause of lung cancer in the United States (after smoking), according to the U.S. surgeon general.

Radon gas finds its way into homes through cracks in floors and walls, floor drains, or construction joints. Radon levels are generally highest in basements and ground-floor rooms. The only way to know the radon level in a building

Map 11. Radon potential in California counties.

Zone 1, highest potential
Zone 2, moderate potential
Zone 3, low potential

is to test the air. Fortunately, simple home test kits are available at low cost, and corrective measures can be taken when high levels are detected. The average concentration of radon in American homes is about 1.3 picocuries per liter (pCi/l), while the average concentration in outdoor air is about

.4 pCi/l. The EPA recommends avoiding long-term exposures to radon above 4 pCi/l. Most counties in central California face a moderate radon risk (map 11). Most of northern California and the state's southernmost counties have a low potential for radon. Only Santa Barbara and Ventura counties show the highest potential for radon risk.

The Enemy Is Us: Emission Sources

Tailpipes

California is the nation's largest auto market, accounting for 12 percent of automobile sales in this country. In 2002, Californians had 28 million registered cars, trucks, and motorcycles, in which we traveled about 810 million miles every day, or about 295 billion miles each year. The residents of this state have earned a worldwide reputation as lovers of cars and freeways. But our infatuation with the automobile has a definite downside: gasoline- and diesel-powered vehicles generate about 70 percent of the air pollution in California (pl. 66).

The gasoline engine mixes about 15 parts of air with every one part of gas. As that air passes through internal combustion engines, heat changes nitrogen in the atmosphere to nitric oxide. Gasoline that is never fully consumed is transformed into a mix of VOCs and particulates, which emerges from tailpipes or as fumes from under hoods.

Achieving reductions in pollution emissions and increases in fuel efficiency has involved decades of fighting with the auto industry. Time and again, the industry has had to be forced to make necessary changes. Repeatedly, auto manufacturers have reacted to proposed reforms by arguing that such measures were technically impossible or that they would bring economic ruin to the nation's autoworkers. Foreign automakers, in many cases, pushed ahead with the needed innovations, at which point Detroit finally followed their leads.

Plate 66. If tailpipes were on the fronts of our cars, would we care more about our personal emissions?

This unfortunate adversarial relationship goes all the way back to the 1930s, when electric trolley lines were being purchased by a company called National City Lines. The company quickly dismantled those mass transit systems in 45 cities across the United States. National City Lines was backed by General Motors, Standard Oil, Phillips Petroleum, Firestone Tire and Rubber, Mack Truck, and other companies attempting to monopolize and promote the sale of buses, automobiles, tires, and oil products. In 1949, a federal grand jury found that a nationwide conspiracy had been under way since the 1930s. The companies behind National City Lines paid minor fines but still achieved their goals.

In San Francisco, 200 General Motors buses replaced a fleet of 180 electric streetcars and 50 electric passenger trains. In southern California, more than 1,000 miles of track and 3,000 Pacific Electric trains stopped serving what had been the world's largest electric mass transit system. Diesel buses and private cars took to a freeway system that kept expanding after World War II.

By 1950, a postwar boom brought the state's population up to 11 million people. That year, there were 4.5 million vehicles registered in California. Severe smog had also become a fact of life. In 1953, Los Angeles County Supervisor Kenneth Hahn wrote the Ford Motor Company asking if the automaker was conducting any research to control exhaust pollution. His reply came from the company's news department: "The Ford engineering staff, although mindful that automobile engines produce exhaust gases, feels these vapors are dissipated in the atmosphere quickly and do not represent an air pollution problem" (Doyle 2000, 22).

California officials kept prodding Detroit in the next decade, with limited success. The first automotive emissions control technology in the nation, positive crankcase ventilation (PCV), was mandated in California for 1963 model cars. The PCV system pulled hydrocarbon gases from engine crankcases to be reburned in the cylinders. The devices eliminated only 25 percent of hydrocarbons in exhaust, however.

A better job of cleaning tailpipe emissions would require catalytic converters, or "afterburners." Catalysts facilitate a chemical reaction without being consumed in the process. Similar technology had been used in coal mining and petroleum industries since the first decades of the twentieth century. Carmakers resisted the use of catalysts until the converters were independently developed outside the automobile industry. After California certified four prototypes in 1964, automakers were required to install these systems on 1966 model cars.

However, American automakers loudly predicted disaster, arguing that the devices could not be built, that they would be far too expensive, and that they might even force a catastrophic shutdown of their industry. They successfully forestalled implementation for years. In contrast, the Japanese automaker, Honda, was the first to announce it was ready to meet the standards. Still, legislators granted the industry a reprieve until 1975. Despite all of the earlier dire predictions, two months after the delay was decided upon, General Motors declared that it could meet the requirement for catalytic converters on all its models after all. Volvo, another foreign automaker, introduced a car in 1977 billed as "smog-free," the first model with a three-way catalyst that worked on nitrogen oxides, hydrocarbons, and carbon monoxide. Detroit automakers did not fully comply with the catalyst requirement until the 1979 model year.

Back in 1969, the U.S. Department of Justice had filed an antitrust lawsuit against the Automobile Manufacturers Association and carmakers American Motors, Chrysler, Ford, and General Motors. What came to be known as the "smog conspiracy case" alleged that the manufacturers had together avoided competition, for at least 16 years, in the development of air pollution control devices. The suit was settled by consent decree and grand jury records were sealed, so much of the basis for the lawsuit was never made public.

The Federal Clean Air Act of 1970 compelled the industry to finally develop such equipment, and it set standards for tailpipe emissions that would reduce pollution by 90 percent from then-current levels. The requirements were "technology-forcing," requiring equipment before it was actually available while setting deadlines to meet those objectives. Public health was the overriding concern. Nonetheless, delays were granted, year after year. Sam Leonard, director of environmental activities at General Motors, would later say, "Ninety-three was the first model year we ever built a model certified to seventy-five standards" (Shnayerson 1996, 50).

After 1973, when the OPEC oil embargo drove up gas prices and limited gasoline availability, a market suddenly developed for smaller, more fuel-efficient automobiles. American automakers faced stiff competition from foreign compact cars like Toyota, Honda, and Datsun. Weight reduction alone, in the smaller cars, accounted for better fuel economy and lower emissions. In the years to follow, much less steel would go into all cars, replaced with plastics that would reduce weight further. Other innovations were mandated in the years to come. Since 1994, cars have had computer systems with "check engine" messages that alert drivers when there are problems in the smog control system.

Considerable progress occurred in controlling auto emissions between 1980 and 1990. That decade saw only a 5 percent increase in fuel consumption. Then progress slowed. Between 1990 and 2003, fuel consumption grew 23.5 percent nationally, as the number of registered vehicles swelled about 22 percent. Significantly, the number of light trucks, especially SUVs, increased by 74 percent in that period.

"Light-duty vehicles" are pickup trucks, minivans, and SUVs that weigh no more than 6,000 pounds. They now account for roughly half of all auto sales in the nation. These vehicles do not have to meet mileage standards mandated for passenger cars and so emit much more pollution. Although the fuel economy of today's light-duty vehicles has improved, on average, by 6 percent since 1987, if horsepower and vehicle weights had not increased so much, fuel economy might have been 58 percent better! Average horsepower has grown by 76 percent, and the average SUV today weighs about 26 percent more than in 1987.

Diesel

The gasoline-powered internal combustion engine generates a spark to ignite fuel compressed with air in a cylinder. Diesel engines compress air much more, making it hot enough to explode without the need of a spark when fuel is injected. It is a more efficient use of energy, per gallon of fuel, but the higher temperatures also cause more nitrogen oxides to form. Some diesel fuel always remains incompletely burned, pumping unhealthy soot particles from truck and tractor exhaust pipes.

While only 2 percent of the vehicles on California's roads are diesel-powered, such vehicles account for 31 percent of the total smog-forming nitrogen oxides and 79 percent of

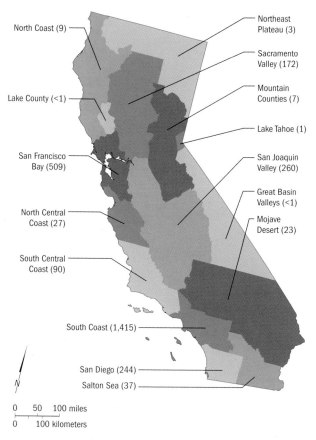

North Coast (9)

Northeast Plateau (3)

Sacramento Valley (172)

Mountain Counties (7)

Lake County (<1)

Lake Tahoe (1)

San Francisco Bay (509)

San Joaquin Valley (260)

Great Basin Valleys (<1)

North Central Coast (27)

Mojave Desert (23)

South Central Coast (90)

South Coast (1,415)

San Diego (244)

Salton Sea (37)

0   50   100 miles
0   100 kilometers

Map 12. Estimated premature deaths due to diesel exposure, by air basin.

the total particulate matter emitted by on-road vehicles. Exhaust from heavy-duty diesel engines contains 100 to 200 times more small particles than gasoline engine exhaust. More than 40 chemicals contained in diesel exhaust are considered toxic air contaminants in California. Diesel exhaust has been listed as a known carcinogen since 1990 (under California's Safe Drinking Water and Toxic Enforcement

Act). The California Air Resources Board estimates that diesel particulate matter contributes to 2,800 premature deaths in California each year, to 3,600 hospital admissions, 240,000 cases of asthma and respiratory symptoms, and 600,000 lost workdays (map 12).

California's truck and bus fleets are required to perform annual inspections of their heavy-duty vehicles. Systems must be properly maintained, must not have been tampered with, and must be free from excessive smoke. Fleets' maintenance and inspection records are randomly audited, and representative samples from vehicles are tested. The California Air Resources Board also operates a roadside smoke inspection program. Heavy-duty diesel-powered vehicles that are not part of a fleet or are exclusively for personal use are exempt from all of this oversight.

A five-minute idling regulation took effect for diesel trucks in 2004. It applied everywhere, except when trucks were being loaded or unloaded or during federally mandated rest periods. Though truckers can now hook up to electricity at many truck stops to keep their air conditioners and heaters running, many drivers and trucking companies have been unhappy about the inconvenience and costs. But research suggests that if the truckers comply with the regulations, 166 tons of soot emissions and 5,200 tons of nitrogen oxides should disappear from California's air each year. Compliance and enforcement will be the keys.

Of the 26,000 school buses in California, about 69 percent of them run on diesel fuel (pl. 67). Because children are more susceptible to the toxicity of air pollution, idling regulations have been specifically created for school buses and other buses or commercial vehicles that operate near schoolchildren. Drivers must turn off their engines when they arrive at any elementary or high school and restart them no more than 30 seconds before departing. School bus drivers are prohibited from idling more than five minutes at school bus stops or at other school activity sites.

Plate 67. A school bus belching toxic diesel exhaust as it accelerates.

The average diesel bus is about 200 times more toxic than a compressed natural gas (CNG) school bus. Though CNG buses cost more, more than 40 school districts in California shifted to them by 2004. Hybrid diesel-electric buses are another improvement, offering a 97 percent reduction in carbon monoxide and 10 percent higher fuel economy.

Ports of Call

Across the United States, millions of pounds of smog precursors enter the atmosphere whenever an aircraft takes off and lands. One 747 jet arriving and departing from Los Angeles International Airport (LAX) produces as many VOCs as a car driven 5,600 miles and as much NO_x as a car driven 26,500 miles (pl. 68). Every day, airports add approximately 16 tons of NO_x and 13 tons of VOCs to the South Coast Air Basin, according to the California Air Resources Board, including emissions from both aircraft and ground equipment.

The EPA has proposed an upgrade to emission standards for NO_x for new commercial aircraft engines. The standards would bring the U.S. standards into alignment with interna-

Plate 68. Jets taking off can generate as much pollution as a car driven thousands of miles.

tional standards set by the United Nations International Civil Aviation Organization. Meanwhile, LAX remains the second-largest industrial source of VOCs and NO_x in the Los Angeles area.

The state's largest fixed source is the Los Angeles–Long Beach port complex. From the coast, ships sailing out near the horizon might present a romantic image, looking small and alone. At a distance, it is difficult to appreciate how massive most commercial cargo ships are today or how many ships are out there on the sea. In fact, ocean-going vessels are major sources of air pollution that have been almost unregulated. Their enormous diesel engines burn the dirtiest form of high-sulfur fuel. They emit nitrogen oxides, particulate matter, sulfur, and toxic air contaminants. For every mile a big container ship travels, it can produce 25,000 times more NO_x emissions than a car does while traveling one mile (pl. 69).

Port operations, like those at Long Beach, San Pedro, and Oakland, add impacts from trucks and trains that service the ships. All these California ports are close to residential neighborhoods, all are expanding, and all present increasing health risks to local communities. In a 2004 report, "Harboring Pollution: The Dirty Truth about U.S. Ports," the Natural

Plate 69. A container ship in the Santa Barbara Channel leaves a long line of diesel soot in its aerial wake.

Resources Defense Council and the Coalition for Clean Air ranked 10 ports by steps taken to reduce impacts to human health and the environment. The Port of Oakland earned the best ranking on this list, a "B– ," while Long Beach received a "C" and Los Angeles (San Pedro) only a "C– ." The two southern California facilities, together, form the third-largest port complex in the world. Every day, 40,000 diesel trucks service them. International trade has grown recently, and the volume of material moving through the complex is expected to quadruple by 2025.

Off Santa Barbara, where there is no commercial port, ships travel 10 to 15 miles from shore. The Santa Barbara County Air Pollution Control District estimates that, even without a port, the offshore ship traffic has negated most of the air quality gains made by reducing car and truck emissions in that county.

Currently available emission-control technologies can cut the pollution from ships by 40 percent or more. And due to the huge size of the ship's engines, the cost-per-ton of reduced emissions is relatively low. Ports can also reduce pollution by providing electric power connections from the shore to docked ships so that auxiliary diesel engines can shut down. In 2002, the Port of Los Angeles announced a program to apply such technology, but implementation has been slow. In 2005, the California Air Resources Board adopted a regulation that requires domestic and foreign ships within 24 miles of the coast to switch to lighter, cleaner diesel fuel. The big question is whether vessels that sail under the flags of other nations will bow to the state's authority that far offshore.

"Air Spray"

Up to 98 percent of the solvent in hair spray evaporates into the atmosphere. A little spray seems so insignificant, so minor an act, yet personal home products like hair sprays have a surprisingly large impact on our air. The long list of volatile consumer products that emit chemicals into the air (sometimes with an odor and sometimes without) includes deodorants, detergents, hair sprays, cosmetics, and insecticides. Their significance is in their cumulative numbers. More than 36 million Californians use half a billion of these items every year.

After tailpipe emissions, personal home products are the second-largest source of smog in southern California. They release nearly three times as many smog-forming compounds as southern California's factories, and five times as many as its gasoline stations. Consumer products put out almost twice as many pollutants as all of California's SUVs and light trucks.

Since 1990, VOCs in these products have been controlled, but regulators face difficulties with enforcement. To fight smog, underarm sprays should all be replaced by bars or roll-on deodorants. However, California law requires that new regulations not eliminate existing forms of products. Manufacturers insist that sprays be considered a separate "product form," though they serve the same purpose as roll-on antiperspirants.

Downwind from the Farm

California boasts the most diverse and productive agricultural system in the world. The multibillion-dollar industry is also a major contributor to the state's air pollution woes. Tractors and irrigation pumps use diesel engines and churn up dust (pls. 70, 71). Farm activities create one-fifth of air pollution emissions in the smoggy San Joaquin Valley. Statewide, agricultural operations represent approximately 14 percent of total diesel particulates emissions.

Plate 70. Exhaust from a diesel irrigation pump in the San Joaquin Valley.

Plate 71. A farm tractor stirs up dust.

In the late nineteenth century, after the citrus industry developed in southern California, orchards were protected from freezing weather using "smudge pots." In these pots, old tires and used motor oil burned, filling the air with black smoke. Some smudge pots were still in use in the 1970s, but they were eventually all replaced with wind machines. Propellers on towers now mix pockets of cold and warm air in orchards.

Until 2004, farms were exempt from emissions rules faced by other commercial enterprises. Pushed by environmental and health interests, though, the EPA has ended the farming exemption. The San Joaquin Valley Air Pollution Control District's new rule in 2004 directed 8,000 farmers to take steps to reduce dust or chemical aerosols (pl. 71). These rules applied to farms with more than 100 acres and dairies with more than 500 mature cows, aiming to cut particulate pollution by 23 percent by 2010 in that valley.

To meet these new standards, farmers can turn to a host of strategies. When possible, they can use machinery to work fields at night, avoiding the heat and wind that contribute to pollution formation and transport. Organic farming methods reduce or eliminate chemical fertilizers and pesticides. In the past, farmers have burned fields to reduce stubble from harvested crops, to dispose of pruned limbs, and to control weeds, but some of this waste material can be ground to mulch instead. Burning will be phased out and then banned in 2010. Using controlled fire as a tool to reduce the hazard of wildfires in foothills and mountain areas will still be allowed, but the air quality districts will coordinate efforts to determine proper burn days.

The number-one dairy producer among all states, California is home to 1.6 million dairy cows and a similar number of young cattle and "dry" cows. Most of the herds are in the San Joaquin Valley, where more than 800,000 cows live in 1,600 dairies. Remarkably, the highly urbanized Chino Basin in Riverside and San Bernardino counties features the highest concentration of dairies anywhere in the world. The 300

Plate 72. Methane, a greenhouse gas, comes from both ends of cows.

dairies there have up to 40 cows per acre. Each of those cows can produce 22 tons of waste in a year. Manure particles that dry out are carried aloft by wind as corralled cattle walk through piles and stir the dust. Ammonium nitrates and sulfates travel downwind from dairies and cattle feedlots. Methane, a greenhouse gas, also emerges from either end of a cow and can total 160 pounds in a year per cow (pl. 72). Whether cows also give off volatile organic gases (methane is organic, but not volatile)—and if so, how much—is still being researched.

Natural Emissions Too

Plants provide us with oxygen, and many of them actually clean pollutants from the air, so how could President Ronald Reagan make the seemingly outlandish declaration, in 1979, that "80 percent of air pollution comes not from chimneys and auto exhaust pipes, but from plants and trees"? The statement actually had some basis in fact, though it overlooked the role that human emissions play when they interact with natural, biogenic emissions. Many trees do release significant amounts of hydrocarbons, particularly when it is hot. The trees let go of some hydrocarbons as a heat control mechanism, akin to the cooling effect of sweat evaporating from your skin.

The Great Smoky Mountains are "smoky" because hardwood forests there generate aerosol particles. California's oak trees emit a type of volatile organic compound called isoprene, as do popular ornamentals like weeping willow and bottlebrush. The familiar, pleasing scent of pine forests is a sure sign that the trees are adding chemicals to the air. Conifers put out several kinds of terpenes. Like other VOCs, isoprene and terpenes can mix with nitrogen dioxide, provided by human activities, and add to the ozone in photochemical smog.

The foothills east of Sacramento receive both NO_x and VOCs blown in from the valley, but that region's oak trees locally provide VOCs for perhaps 70 percent of the ozone that ultimately forms there on hot days, in reactions with NO_x from human sources. Emissions from pine trees, on the other hand, actually consume ozone, but convert it to aerosols that make the sky hazy. It is important to remember that, without vehicle exhaust products first contributed by us, tree emissions themselves would not be a big issue.

Sharing Air with the Wildlands

Most of us eagerly anticipate long-range vistas and clear air on a visit to a national park. Such special places are meant to be wild, clean, and protected. However, air is no longer pristine in many of our wildlands. In 1999, the EPA announced an effort to improve air quality in national parks and wilderness areas. The "Regional Haze Rule" aims to improve visibility at 29 sites in California, requiring the state and federal land management agencies to reduce the pollution that impairs visibility (map 13). Particulate haze that scatters and absorbs light is the primary target of this effort.

Map 13. National parks, national wilderness areas, and national monuments with special air quality protections.

Sequoia and Kings Canyon National Parks: A Case Study

Located downwind from the San Joaquin Valley, Sequoia and Kings Canyon national parks experience some of the worst air quality of any of the nation's national parks. The air in western portions of these adjoining parks can be as polluted as anywhere in the South Coast Air Basin, on any given day. Air quality has been studied and monitored at Sequoia and Kings Canyon for almost 20 years. The parks provide input in state and federal regulatory processes to shape decisions that benefit the natural resources they are charged to protect. Their efforts include a concern for the health and experiences of park visitors (and of park employees). Within the parks, specialists monitor the atmospheric transport of ozone, acids, pesticides, and other synthetic chemicals into the parks. Lost visibility due to haze and loss of the naturally dark night sky due to light pollution are subjects of scrutiny.

On a clear day, from the western edge of Sequoia or Kings Canyon national parks, the Coast Range is visible 100 miles away (pl. 73). Few days are that clear now. Instead, visitors often look out over a pall of haze trapped in the inversion layer of the San Joaquin Valley.

Nearly 90 percent of the Jeffrey pines *(Pinus jeffreyi)* in or near the Giant Forest portion of Sequoia national park have shown visible signs of ozone injury, similar to that seen in the seriously impacted forests in southern California (pl. 74). Ozone-weakened trees lose some of their ability to fight off bark beetles and the effects of drought. Damaged by pollution, their chlorophyll (the green pigment in needles) cannot carry on photosynthesis at the same rate. Needles drop earlier than normal, and trees show less annual growth. Fortunately, the massive old giant sequoias *(Sequoiadendron giganteum)* seem relatively resistant to ozone levels, but sequoia seedlings are vulnerable to ozone injury.

Acidity and alkalinity are measured on the pH scale.

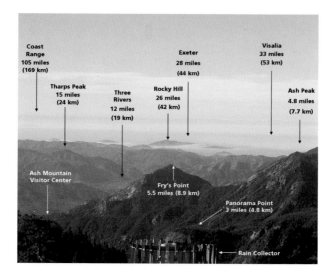

Coast
Range
105 miles
(169 km)

Tharps Peak
15 miles
(24 km)

Three
Rivers
12 miles
(19 km)

Rocky Hill
26 miles
(42 km)

Exeter
28 miles
(44 km)

Visalia
33 miles
(53 km)

Ash Peak
4.8 miles
(7.7 km)

Ash Mountain
Visitor Center

Fry's Point
5.5 miles (8.9 km)

Panorama Point
3 miles (4.8 km)

Rain Collector

Plate 73. *Top:* A good air day, with visibility from Sequoia National Park all the way to the Coast Range. *Above:* The same view on one of the increasingly common bad air days.

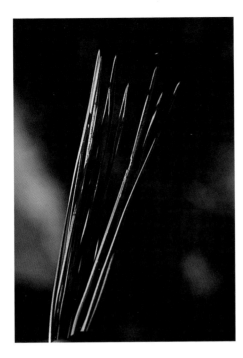

Plate 74. Visible damage to the chlorophyll in pine needles is caused by ozone.

"Neutral" on this scale is represented by the number 7; pure distilled water establishes that standard. Clean rain is naturally slightly acidic, because water vapor combines with carbon dioxide in the air to make carbonic acid. Most rain measures about pH 5.6. Since each whole number on the scale represents a change by a factor of 10, a pH of 5 is 10 times as acidic as pH 6 and 100 times more acidic than pure water.

Acid rain may form when nitrogen oxides, sulfur dioxide, and chloride ions enter the air. It arrives in the Sierra Nevada in rain or snow, or simply drops out of the sky as dry particles. Natural systems evolved and adapted to the normally weak acidic condition of rain, but elevated acid levels can cause extensive damage to forests and to life in aquatic ecosystems. When the pH drops to between 5 and 4.5, trout begin to suffer reproductive problems. In waters below pH 4.5, adult trout die.

Such effects are more common in the northeastern United States, where sulfur dioxide from coal power plants generates acid rain. During California's winter wet season, however, acids can be deposited in the Sierra Nevada snowpack with each storm. They accumulate there until snow melts in the spring. Acidic water suddenly runs off to mountain lakes and streams, just as many aquatic creatures are beginning their annual reproductive cycles. In the Sierra, ongoing levels of acid deposition that are low, compared to those in the east, may eventually turn into big pulses of concentrated acid.

Similarly, melting snow contaminated by pesticides can also dump these contaminants into water systems. Every year, tons of pesticides are applied to crops in the San Joaquin Valley, upwind from the Sierra Nevada national parks. Pesticides and organophosphates from fertilizers drift on the prevailing west winds into the mountains. Synthetic chemicals have been detected in precipitation as high as 6,300 feet in Sequoia National Park.

The Foothill Yellow-legged Frog *(Rana boylii)* completely disappeared from Sequoia, Kings Canyon, and Yosemite national parks in the 1970s and today persists in the Sierra Nevada only in a few places scattered across the foothills. There have also been major population declines in Red-legged Frogs *(R. draytonii),* Yosemite Toads *(Bufo canorus),* and Mountain Yellow-legged Frogs *(R. muscosa)* (pl. 75). The Red-legged Frog is listed as threatened under the U.S. Endangered Species Act, and the other two amphibians have been proposed for listing.

Though the introduction of nonnative predatory fish probably explains much of the lost frog population, research also suggests that pesticides, blown from agricultural areas, are affecting amphibians in mountain ponds and streams. By contrast, healthy numbers of frogs still live in the Coast Ranges, west of the San Joaquin Valley, upwind from most

Plate 75. Mountain Yellow-legged Frogs.

drifting pesticides. One study, published in 2002, found a strong correlation between declines in four species of frogs and living in the path of winds coming from Central Valley farm lands (map 14). (Alternate possibilities that such declines might be due to global warming or higher ultraviolet radiation levels were not demonstrated in that study.)

Scientists from the U.S. Geological Survey (2000) collected tadpoles and adult frogs from coastal sites and from the Sierra Nevada foothills, the Tahoe Basin, and Yosemite, Sequoia, Kings Canyon, and Lassen national parks. Some pesticides absorbed by frogs suppress a nervous-system enzyme, cholinesterase. Researchers found tadpole populations in the mountains with lower overall values for the enzyme than those populations along the coast. Within the Sierra Nevada, enzyme activity levels were significantly lower in tadpoles in the southern parks, east of the San Joaquin Valley, than they were in similar sites farther north, where there is less agricultural activity. More than half of the adult frogs and tadpoles at Yosemite National Park showed measurable levels of two common pesticides, chlorpyrifos and diazinon. On the coast, only 9 percent of the animals did.

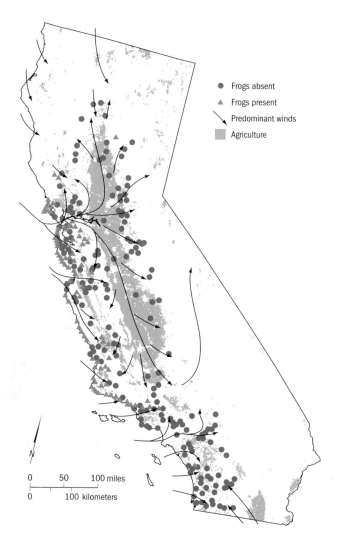

Legend:

● Frogs absent
▲ Frogs present
↗ Predominant winds
▨ Agriculture

Map 14. Patterns of decline for red-legged frogs, the distribution of agricultural lands, and predominant wind directions.

In the High Sierra, the air is commonly very transparent. Amateur astronomers often come to national parks, far from city lights, to view the stars typically visible on a clear night. Today, however, when stargazers look west at night from Sierra Nevada vantage points, a continuous glare stretches from Sacramento to Bakersfield (map 15). The trespassing light means that evidence of human technology is always there to intrude on the landscape. Dark skies full of visible stars shaped our history and cultural evolution; now they must be added to the list of increasingly rare resources deserving protection in our mountain parks.

Down in the lowlands, where the lights are concentrated, city residents can generally see no more than a few dozen of the brightest stars. Constant twilight substitutes for true darkness, and many people's eyes never fully adapt to darkness anymore. Nearly two-thirds of the world's population now lives under night skies polluted by light. Truly dark, starry skies are unavailable to 99 percent of the people in the continental United States.

Epidemiologists trying to understand why breast cancer rates in industrialized societies have been five times higher than in developing countries, have begun focusing on light pollution. They're responding to research that showed a 20 to 50 percent lower risk of breast cancer in blind women. Light inhibits the nighttime secretion of melatonin in our brains. That hormone triggers a reduction in the body's nocturnal production of estrogen. A streetlight shining on a shaded window may send enough light into a bedroom to cause this effect. Long-term decreases in melatonin secretion may be tied not only to breast cancer, but also to other estrogen-related diseases, including prostate cancer, depression, chronic fatigue, and reproductive anomalies.

Truly dark skies could be recovered in many places. Lights can be shielded and focused with reflectors to reduce "light trespass." Several communities in California have adopted light-reducing ordinances to curb light pollution through

Map 15. City lights across California at night.

better shielding and light efficiency. Ideally, all outdoor lighting would be shielded. With "flat glass fixtures," streetlights near homes can focus their light more toward the street. Homeowners can install shades to block light coming through bedroom windows. Of course, outside lights may also simply be turned off.

While California attempts to control air pollution within its own borders, the atmosphere keeps circulating, sharing air and the pollution it carries, around the globe.

The atmosphere remembers our past behavior. There are limits to the compassion of the forgiving air.

—RICHARD SOMERVILLE (1998, 30)

Adrift from Asia

Winds can cross 5,000 miles of the Pacific Ocean and make it to the California coast in less than 10 days. At Arcata, on the north coast of the state, instruments detect the arrival of dust and pollution that originates in Asia. In recent years, ozone levels reaching California across the Pacific Ocean have been 30 percent higher than levels detected in 1985.

Formation of an "atmospheric brown cloud" has become an annual spring phenomenon over much of eastern Asia. Springtime dust storms in central China pick up smog and particulate pollution and carry it aloft. The phenomenon has increased because Asian economies have been rapidly modernizing. Consumers in China, in particular, are abandoning bicycles as modes of transportation and buying cars instead. Most of the automobiles sold in China (many by American carmakers) do not meet the current U.S. standards for air pollution control. The Chinese are also heating bigger homes, mostly with coal. The severe air pollution that results only adds to California's air quality challenges.

The world's industrial and leisure activities have also created a persistent layer of pollution and dust, several miles thick, that hovers about three miles above the Earth's surface. That layer seems to contribute to global darkening on the ground, where measurable sunlight has decreased up to 20 percent at times. High-altitude air pollution may also be changing rainfall patterns. Aerosols normally serve as nuclei for raindrop formation, but when particles are too tiny, the water that gathers around each nucleus never becomes heavy enough to form a drop and fall. More and thicker clouds may appear over regions polluted with fine particulates, but these never produce local rain. Of course, water vapor must eventually cycle back to the surface as rain or snow, but small aerosols may be shifting rainfall away from urban regions, where demand for water is greatest.

Ozone Depletion

Ozone near the ground damages human health and the environment, but the layer of ozone up in the stratosphere is essential to life, shielding us from ultraviolet radiation (as detailed in The Thin Blanket). Researchers in the Antarctic detected thinning of the ozone shield in 1981, and satellite observations eventually confirmed their data. Because the effect has been most pronounced over the poles, some of the media began using the term "ozone hole." More accurately, there has been a general thinning of the stratospheric layer due to destruction of ozone, at an average rate of about 3 percent per decade from 1978 to 1991. But this loss has not been uniform across the globe. Ozone has diminished most at the poles, while the tropics have experienced almost no change. Over the United States, the ozone layer declined by 4 percent per decade from 1978 to 1991.

During this period of significant loss, the culprits were man-made chemicals that found their way up to the stratosphere and catalyzed reactions that broke apart ozone. Chlorofluorocarbons (CFCs) had been used as refrigerants, insulating foams, solvents, and the propellants in spray cans. Millions of tons of them entered our atmosphere. When CFCs are broken down by sunlight, they release chlorine, which breaks ozone apart. Each chlorine atom repeatedly combines with as many as 100,000 ozone molecules during its long stratospheric life.

Other ozone-depleting substances include pesticides such as methyl bromide, halons used in fire extinguishers, and methyl chloroform used in industrial processes.

After the thinning of the ozone shield became evident, the world came together to correct the problem. The United States took a leadership role in securing adoption of the Montreal Protocol in 1987, which was ratified by 57 countries. These nations agreed to phase out CFCs and halons. When additional scientific evidence confirmed that the ozone loss was caused by CFCs, 120 countries met with a sense of urgency in London in 1990 and strengthened the Protocol agreement. They accelerated the timetable for the phase-out and made 2000 the new deadline. Meanwhile, economically viable alternatives for CFCs became available, which contributed to the success of this international effort. Developed nations contributed to an international fund to assist poorer, undeveloped nations in making the shift to the new substitutes.

Millions of tons of persistent CFCs and other ozone-depleting chemicals are still up there, in the stratosphere, and it will take decades for them to break down so the ozone layer can be naturally restored. There is good news, however. Although ozone depletion is still occurring, the rate has slowed considerably and is expected to peak soon. It may take 50 years for natural processes to bring back the upper atmosphere's ozone layer. Until that time, we will experience increased levels of ultraviolet (UV) radiation at the Earth's surface.

In response to the increased exposure risks, the National Weather Service and the EPA have developed a UV index, computed daily and generally broadcast as part of weather forecasts (table 6). A computer program generates a number, on a scale of zero to 10, based on ultraviolet levels forecast for noon (when exposure peaks) and taking into account the time of year and latitude (to factor in the position of the sun on that date). Adjustments are made for local cloud cover: in a sky with only scattered clouds, 89 percent of UV reaches Earth; broken clouds let 73 percent through; and overcast conditions can still allow 31 percent of UV to reach the ground. You can be sunburned on an overcast day. Altitude is another local factor. Exposure increases about 6 percent for every 3,000 feet above sea level, so spring skiers at 9,000 feet on Mammoth Mountain are experiencing 18 percent more UV exposure than swimmers at Newport Beach.

People with very sensitive skin and infants always need protection from prolonged sun exposure. Hats and sunglasses help protect eyes from UV damage. Sunscreens,

TABLE 6. UV Index

Index Values	Air Quality Description (Color)	Health Cautionary Statement
0-2: Minimal	Blue	Minimal danger from the sun's UV rays for the average person. Most people can stay in the sun for up to 1 hour during the peak sun strength, 10 a.m. to 4 p.m.
3-4: Low	Green	Low risk of harm from unprotected sun exposure. Fair-skinned people, however, might burn in less than 20 minutes.
5-6: Moderate	Yellow	Moderate risk of harm. Fair-skinned people might burn in less than 15 minutes.
7-9: High	Orange	High risk of harm. Fair-skinned people might burn in less than 10 minutes. Minimize sun exposure during midday hours, from 10 a.m. to 4 p.m.
10+: Very High	Red	Very high risk of harm from unprotected sun exposure. Fair-skinned people might burn in less than 5 minutes. Minimize sun exposure during midday hours. Avoid being in the sun as much as possible.

Modified from www.epa.gov/sunwise/stayheal.html.

liberally applied, should have a "sun protection factor" (SPF) of at least 15.

CFCs not only deplete the ozone layer; they are also "greenhouse gases," which makes controlling them all the more critical.

The Enhanced Greenhouse Effect and Climate Change

Gambling with Greenhouse Gases

Across the globe, researchers, environmental advocates, and increasing numbers of the general public share one pressing concern about the Earth's atmosphere in the twenty-first century: the gamble we are taking with greenhouse gases. About six billion tons of carbon, one ton for every person on Earth, are added to the atmosphere each year. Atmospheric carbon dioxide, mostly from the burning of fossil fuels, has been increasing by about .4 percent per year. Since the year 1800, the level has risen from about 270 parts per million to over 370 parts per million. That number is almost certain to grow past 540 parts per million, doubling the preindustrial level, sometime near the middle of this century. Despite global concerns, there has not yet been any significant change in the rate of increase, so levels could be much higher by the end of this century. Already, carbon dioxide concentrations are greater than anything seen in the last 650,000 years (Siegenthaler et al. 2005).

Other greenhouse gases are increasing too. Carbon dioxide is only part of this problem. Methane, nitrogen oxides, ozone, and manmade chemicals like CFCs are present in smaller amounts, but each is much more efficient than carbon dioxide as a greenhouse gas. Levels of methane, in particular, are 130 percent higher than they have been in the past 650,000 years.

Partly due to this enhancement of the natural greenhouse effect (as discussed in The Thin Blanket), global temperatures are rising (fig. 20). The warming trend has been most rapid during the second half of the twentieth century. Ten of the warmest years of the last century have occurred in the last 15 years; 1998, 2002, and 2003 were the three warmest. When we compare decades, the 1990s appear to be the warmest since the fourteenth century.

The Earth's climate has always varied (fig. 21). In the fourteenth century, the globe entered the so-called Little Ice Age, in one of its periodic swings toward cooling. Ice ages have alternated with warmer periods repeatedly. In just the last decade, climate scientists have developed new techniques that document how much change has characterized the Earth's climate. Annual layers in cores from ice shields on Greenland and the Antarctic preserve a climate record much like annual

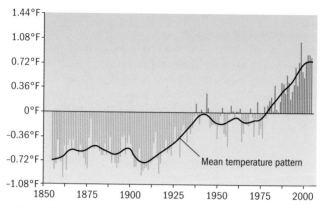

Figure 20. Global surface temperatures (combined land and sea surface), 1856–2003.

growth rings on trees. The cores span 650,000 years and record temperature and carbon dioxide and methane conditions when the ice was deposited. At other sites, like the bed of Owens Lake, sediments tell a story spanning 800,000 years of deposits. Accumulating evidence from deep-sea zones gives an accurate record of temperatures for the past 40 million years! Across time, carbon dioxide and temperature have moved in correlated patterns. Whenever solar cycles, variations in the Earth's orbit, and natural global warming reached some trigger point, warmer climates abruptly shifted into ice ages.

Climate change is a natural phenomenon. What humans added to this story, beginning perhaps 8,000 years ago, after agriculture developed, and continuing once the industrial age got under way, was an increasing infusion of greenhouse gases. Flood-irrigated farm fields brought about plant decomposition and rising levels of methane. Industrial burning and deforestation released far more carbon dioxide. In the last four decades of the twentieth century, when the human population more than doubled (from three billion in 1960 to about 6.5 billion in 2005), human activities became more

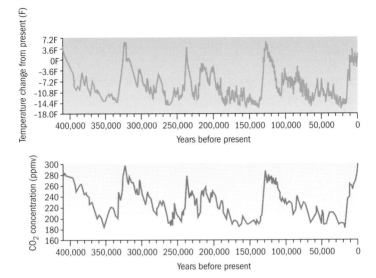

Figure 21. Temperature and carbon dioxide concentrations in the atmosphere over the last 400,000 years (from ice cores).

clearly responsible for the magnitude and the speed of changes. Growing evidence shows that, in our massive numbers and with our omnipresent technologies, we are skewing the globe's natural climate variability.

Plants in the oceans and on land pull carbon dioxide out of the air as they carry on photosynthesis. But in recent decades, so much deforestation has occurred (with associated burning and decomposition) that, overall, land plants have contributed to an increase in carbon dioxide releases into the atmosphere. Ocean plants have been more effective at countering the carbon dioxide buildup. About 48 percent of manmade carbon dioxide has been sequestered in the seas. Oceans additionally moderate temperatures by absorbing so much heat from the warmer atmosphere. Warmed seawater mostly lies near the surface so far, but slow mixing of the heated water has been detected 10,000 feet deep.

Some skeptics about the "reality" of enhanced global warming have focused on a supposed disagreement between land-based and satellite-based temperature data. An early report suggested that satellite instruments were not detecting an atmospheric temperature increase. Independent reanalysis of those data located its methodological error. When corrections were made, no actual discrepancies persisted between the two sources of data.

"Global warming" has become the catchphrase used to summarize a range of climate change issues. But an array of other symptoms besides temperature serves as evidence for the enhanced greenhouse effect. Sea levels rose about four to eight inches in the twentieth century. Melting water runs from land into the sea when glaciers and the Antarctic ice caps melt, but most of the rise is actually due to thermal expansion as the ocean gets warmer. Since the arctic ice is not on land but already floating in the sea, it does not contribute anything additional to the sea level when it melts (although ocean-cooling and dilution effects are a concern, as discussed in Slow or Sudden).

Along with rising sea levels, a warmer ocean and atmosphere generate more severe weather and storm patterns. Researchers with the Scripps Institute of Oceanography in San Diego have documented rising trends in storm wave heights and the incidence of "extreme waves" during the last 20 years.

High overhead, something else has been rising. The boundary between the troposphere, the lowest layer of the atmosphere, and the stratosphere has moved to a higher elevation in the past two decades. High-altitude weather balloons documented the shift. Scientists at the Lawrence Livermore Laboratory explained the phenomenon (using weather forecasting models) as the result of a cooler stratosphere, due to ozone depletion, and a warmer troposphere, due to the increase in greenhouse gases.

Evidence consistent with global climate change is already visible in California's plant and animal communities. In the

Plate 76. Edith's Checkerspot Butterfly at Mineral King in the southern Sierra Nevada.

last century, the Edith's Checkerspot Butterfly *(Euphydryas editha)* shifted its range northward and upward, to follow the cooler conditions it required (pl. 76). Wildlife surveys in Yosemite in 2003 found Piñon Deer Mice *(Peromyscus truei)* up to 10,240 feet—a significant change from their earlier range, documented in a meticulous survey 80 years ago that found them only below 9,000 feet. American Pika *(Ochotona princeps)*, the small rodents that stack harvested grass into winter "haypiles," could no longer be found in parts of the park where they had been common (like the Glen Aulin High Sierra Camp); they had also shifted to higher elevations.

The warming ocean current off the California coast has coincided with population shifts of aquatic organisms. In Monterey Bay, southern species have increased, while the numbers of northern species of zooplankton and at least one seabird, the Sooty Shearwater *(Puffinus griseus)*, have declined. Researchers have noted fewer northern, cold-water fish species in the kelp forests off the southern California coast. For example, Green-spotted Rockfish *(Sebastes chlorostictus)* declined by half after the 1970s. In the same period, populations of warm-water fish, like Garibaldi *(Hypsypops rubicundus)*, increased significantly.

The spring melt of the Sierra Nevada snowpack has shifted, coming earlier in the year now than it has before. The effect has been more prominent in the northern and central Sierra, where peaks are lower, than in the southern parts of the range. Many of the Sierra Nevada's small glaciers are disappearing or are already gone.

The overall weight of such evidence is compelling. Scientists have reached an international consensus about the reality of the enhanced greenhouse effect and humanity's role in driving that phenomenon. Established in 1988 by the World Meteorological Organization and the United Nations Environment Program, the Intergovernmental Panel on Climate Change (IPCC) issued its Third Assessment Report in 2001, the product of several working groups involving 2,500 scientists. The latest assessment included these conclusions: "There is new and stronger evidence that most of the warming observed over the last 50 years is attributable to human activities. Global average temperature and sea level are projected to rise under all . . . scenarios. The projected rate of warming is much larger than the observed changes during the 20th century and is very likely to be without precedent during at least the last 10,000 years" (IPCC 2001, 10–13).

The American Geophysical Union also issued a position statement, in December 2003, that was adopted unanimously by the group's governing council to represent the consensus of the 41,000-member organization:

> Human activities are increasingly altering the Earth's climate. . . . Natural influences cannot explain the rapid increase in global near-surface temperatures observed during the second half of the 20th century. It is virtually certain that increasing atmospheric concentrations of carbon dioxide and other greenhouse gases will cause global surface climate to be warmer . . . a real basis for concern. (American Geophysical Union 2003, 1)

In 2001, the president of the United States asked the National Academy of Sciences to analyze the research and conclusions of the IPCC and decide whether concern about global warming was justified. The academy's report, to an administration that was publicly skeptical about this issue, included this statement:

> Greenhouse gases are accumulating in Earth's atmosphere as a result of human activities, causing surface air temperatures and subsurface ocean temperatures to rise. Human-induced warming and associated sea level rises are expected to continue through the 21st century. The committee generally agrees with the assessment of human-caused climate change presented in the IPCC Working Group I (WGI) scientific report, but seeks here to articulate more clearly the level of confidence that can be ascribed to those assessments and the caveats that need to be attached to them. (National Academy of Sciences 2001, 1)

Slow or Sudden?

"Climate change" is a better description for the current state of affairs than "global warming" because we have seen, throughout Earth's history, that effects of warming lead eventually to major cooling cycles in the global climate. The counterintuitive outcome happens because a current, called the ocean conveyor belt, travels through the world's seas, driven by balanced forces of heat and salinity (fig. 22). In a pattern similar to atmospheric wind cells, water in the high latitudes of the globe cools and sinks, while warmer water from the tropics travels nearer the surface. Meanwhile, freshwater entering the ocean from polar ice caps dilutes the ocean's salinity, while evaporation near the equator concentrates salts in tropical water. Thus, a "thermohaline" conveyor begins moving around the world, trying to equalize these differences in temperature and salinity. Conveyance of water through the ocean basins eventually delivers Gulf Stream water to the coasts of Europe and Great Britain, which warms

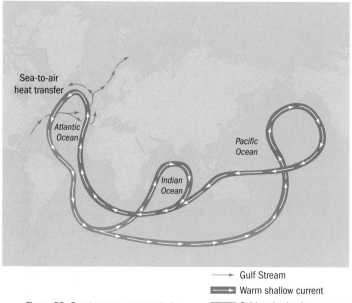

Figure 22. Great ocean conveyor belt.

⟶ Gulf Stream
▰▰▰▰▶ Warm shallow current
▰▰▰▰▶ Cold and salty deep current

those northern climates. A melting arctic ice cap, due to global warming, could send enough freshwater into the northern seas to abruptly shut down the conveyor.

It has happened, repeatedly, in the past. When the conveyor shut down 8,200 years ago, Europe experienced a century of cold, severe winters; glaciers advanced and rivers froze. Around 12,700 years ago, a similar event lasted for about 1,300 years! Icebergs floated in the Atlantic Ocean as far south as Portugal.

A public report prepared for the Pentagon by the Global Business Network envisioned a scenario based on an abrupt climate change of this type. In this scenario, temperatures dropped 5 to 10 degrees F in a single decade in Asia, North America, and northern Europe. Harsher winter weather and reduced moisture for crops in some areas led to a decrease in the number of human lives Earth could sustain.

The Department of Defense was most interested in what happened next in this scenario, the political instability that might follow from such chilling climate change. With food and water shortages, tensions between nations could lead to wars. The authors of the report opened with a statement that their scenario might be extreme, yet they considered it a plausible possible future that "would challenge United States national security in ways that should be considered immediately" (Schwartz and Randall 2004, 1).

Reacting to Change

The nations of the world have taken tentative steps to address the global greenhouse gamble. In Kyoto, Japan, in 1997, the United States helped craft an international commitment to reduce greenhouse gas emissions to 7 percent below 1990 levels by 2012. Although the Kyoto treaty took effect in February 2005, with 141 nations participating, it was never ratified by the United States Senate (the only other industrialized nations not participating were Australia and Monaco). The United States, instead, developed a "domestic climate-change strategy" with voluntary measures and no caps on carbon dioxide emissions. North America remains the highest fossil-fuel-burning, CO_2-emitting region of the world, with annual emissions continuing to increase, rather than decline.

Why have international and domestic efforts been so ineffectual? Many political and business leaders have attributed their foot-dragging (or their outright opposition) on these issues to economic concerns and a wish to sustain the comfortable status quo. In this country, preserving the "American way of life" has become an argument against measures to address climate change. In this vein, some of those in power have criticized atmospheric models used to forecast future conditions. Models generally include calculations and descriptions of levels of certainty for different variables. Some skeptics seemed to focus on the existence of uncertainty, at any level, as reason to discount predictions generated by modelers.

The issue of economic costs rests on whose way of life,

specifically, is at risk from change. Oil companies, for example, have vested interests in the current economy based on fossil fuel consumption. In 1999, a California legislative panel met to consider "Climate Change: Its Economic and Environmental Effects." The panel concluded that reducing global warming emissions in California would benefit not only California's environment and public health, but also the state's economy. The conversion to new technology would generate high-wage jobs, expand clean industries, bring about new markets for California products, and reduce costs associated with imported oil and fossil fuels (California Legislature 1999, 78). An independent analysis released in January 2006 also concluded that economic worries were unfounded. The Center for Clean Air Policy found that California's 2020 emissions goals could not only be achieved at no cost to consumers, but also would ultimately save them money.

Some major corporations have taken positive stands on these topics. The opportunities for new markets and cost savings have prompted public commitments by DuPont, 3M, Toyota, Compaq, Xerox, Maytag, and Shell to reduce energy consumption and emissions.

California and Climate Change

"Think globally, act locally," one saying goes. Global climate change translates into regional and local concerns across California. The state is likely to see average annual temperatures rise by 4 to 6 degrees F in the next century, even assuming actions *are* taken to reduce emissions of greenhouse gases. If no such changes are made, a "higher-emissions scenario" projects statewide temperature averages in California 7 to 10.5 degrees F higher. The range of figures comes from two models whose projections were summarized by the Union of Concerned Scientists in 2004.

Since a warmer climate evaporates more water, more precipitation may fall, but as rain instead of winter snow. The state may also see increased winter runoff, instead of the classic late-spring and summer runoff periods. This change in the typical pattern would reduce flows in rivers in the late spring and summer, but increase the danger of winter floods. The Department of Water Resources estimates that a 3-degree temperature increase could mean an 11 percent decrease in annual water supply. Under the coolest climate change projections, there will be a loss of about five million acre-feet/year in snowpack water.

If more thunderstorms form over land, due to more moisture in the warm air, that might increase the delivery of acids and pollutants to Sierra Nevada lakes. Cloudiness cuts both ways, however. Clouds can either cool an area by blocking sunlight or keep it warm, functioning as a blanket in cold weather. No one knows for sure how the effects of clouds might play out.

California's agriculture uses 80 percent of the state's developed water supply. Many of the crops grown in the state are water-intensive, including alfalfa, cotton, rice, and grapes. Those crops, in particular, might be vulnerable to a less certain water supply.

The sea level has already risen four to eight inches in the last century. It may rise another six inches in the next few decades. Saltwater will move farther into the Sacramento–San Joaquin delta (pl. 77). Since the delta is the key point where pumps pull freshwater into aqueducts sending water to southern California, the water supply for millions of people and farms in the San Joaquin Valley may be jeopardized. This change will also impact fish and invertebrates using the bay/delta waters as a nursery.

A hotter climate will exacerbate air pollution. Heat helps photochemical smog to form. It also causes more fuel to evaporate, makes engines work less efficiently, and increases electric power demands as air conditioners click on.

The California Energy Commission has a climate change web page that summarizes the state's concerns. The opening sentence on that page reads: "Global climate change is real."

Plate 77. Islands in the Sacramento–San Joaquin delta are below sea level, protected by levees, but at increasing risk when the sea level rises.

The state's "water supply, agriculture, forestry, energy production, health, transportation, tourism" will all be challenged, according to the Commission.

Climate Stewardship

The choices Californians make about climate change will have long-range effects. They are important not just now, but for this generation's grandchildren and great-grandchildren. Consequences of decisions made in the first half of this century will not appear until the second half of the century.

Sequestering carbon has been suggested as one response to global warming. Growing plants take up carbon and store it until they die and decompose or are burned. New or replacement forests might be planted, with many benefits, but an area the size of Australia would have to be reforested to keep up with modern carbon output. Then, somehow, those trees would have to be prevented from ever burning or decomposing and releasing their stored carbon. As a short-term response to this crisis, sequestration has benefits. As a long-term solution, it is totally ineffective.

There have even been proposals, and some initial experiments, to fertilize algae using iron and thereby promote more sequestration of carbon in the oceans. But the risks of unintended and unpredictable consequences to ocean ecosystems seem very high.

Some Californians might think that proposals to reduce the use of fossil fuel are just as farfetched. Yet reducing fossil fuel subsidies may be a good and effective first step. Significantly increasing the gasoline tax and applying the revenue to cleaner alternatives to the internal combustion engine is another policy option, but one that seems very unlikely in today's political climate.

Some good news was hiding in the defeat of the Climate Stewardship Act in the U.S. Senate in 2003: the vote was 43 to 55. That the bipartisan measure, sponsored by Senators McCain and Lieberman, received so many "yes" votes was encouraging. Less surprising was that some opponents to that measure called global warming a myth and a hoax.

The legislation did not fade away after that initial defeat. Ten Republicans and 10 Democrats teamed up in 2004 to introduce the Climate Stewardship Act in the U.S. House of Representatives. The proposal would have set mandatory, but modest, greenhouse gas pollution reductions. All major sectors of the nation's economy would have limited greenhouse gas pollution to year 2000 levels by 2010. Companies would have been encouraged to make investments in energy-efficient technologies and renewable energy. That effort also failed but perhaps, by the time this book is published, the logjam of U.S. inaction will finally be broken.

Meanwhile, governors of the states of Washington, Oregon, and California have concluded that "global warming will have serious adverse consequences on the economy, health, and environment of their states, and that the states must act individually and regionally to reduce greenhouse gas emissions and to achieve the benefits of lower dependence on fossil fuels" (California Energy Commission 2004). They formed the West Coast Governors' Climate Change Initiative to develop strategies that require regional cooperation and action. Their list of topic areas includes hybrid vehicles, diesel emissions from ports, renewable energy, energy efficiency, hydrogen fuel, and the development of greenhouse gas monitoring and inventory systems.

California's governor also made a personal commitment to convert one of his fleet of gas-guzzling Hummer SUVs into a hydrogen-fueled vehicle (pl. 78). More significantly, he issued an executive order in 2005 establishing greenhouse gas reduction targets: by 2010, reductions to 2000 emissions levels; by 2020, reductions to 1990 levels; and by 2050, reductions to 80 percent below 1990 levels. The secretary of the California Environmental Protection Agency will coordinate the effort throughout state agencies.

Across the nation, including in California, many cities also decided they could not wait for federal leadership on this issue. Instead, they joined "Cities for Climate Protection," an effort coordinated by the International Council of Local Environmental Initiatives. In 2006, the California cities and counties enrolled in the effort included Arcata, Berkeley, Chula Vista, Cloverdale, Cotati, Davis, Fairfax, Healdsburg, Los Angeles, Marin County, Marin Municipal Water District, Novato, Oakland, Petaluma, Rohnert Park, Sacramento, San Anselmo, San Diego, San Francisco, San Jose, Santa Clara County, Santa Cruz, Santa Monica, Santa Rosa, Sausalito, Sebastopol, Sonoma, Sonoma County, West Hollywood, and Windsor. In San Francisco in 2004, a climate action plan issued by the City aimed to cut greenhouse emissions to 20 percent lower than 1990 levels by the year 2012. The plan relied on incentives to coax people to use buses or bicycles, called for the conversion of many city buildings and vehicles to "green

Plate 78. A prototype hydrogen-powered Hummer and Governor Schwarzenegger at the 2004 opening of a hydrogen fueling station at LAX.

power," promoted energy-saving designs for buildings, and encouraged recycling.

California's carbon dioxide emissions, per person, are about 40 percent lower than the national average. That figure is partly explained by the state's milder climate (which lowers the demand for artificial heat sources), but California has also been a leader in energy policies that promote renewable energy and cleaner air. State law requires that the percentage of renewable energy in the state's electricity generation mix be expanded by 1 percent each year. Utility bills in the state now include a "public goods charge" that pays for research and development on climate change.

In 2002, California took another leadership step, in keeping with its long and innovative history of air quality control, by enacting the "Automotive Greenhouse Gas Law." With transportation accounting for about 70 percent of carbon

Figure 23. Cartoon in the *Ventura County Star*.

dioxide emissions in California, the California Air Resources Board (CARB) was directed to develop regulations to achieve feasible, cost-effective reductions in greenhouse gas emissions from new cars and trucks, beginning in 2009. CARB is now no longer focused only on ground-level air pollution.

To address the worries of those who feared economic consequences from such changes, the new regulations could not limit vehicle choices or have significant effects on the state's economy or driving patterns. When CARB adopted the new regulations, in September 2004, they estimated that the rules would add about $325 to the price of a new car, truck, or SUV in 2012. During the second phase, for model years 2013 through 2016, average costs would increase by about $1,050. But lower operating expenses due to better efficiency would mean overall savings for vehicle buyers, according to CARB. "Responding to climate change need not be an additional burden but can, in fact, promote economic development, ensure energy and economic security, and improve public health and safety" (CARB 2004c).

Under the regulations, by 2016, automakers would be required to reduce greenhouse gas emissions by 34 percent for smaller vehicles and 25 percent for large trucks and SUVs. Technology that was already in place on some automobiles could be used to meet the new standards, such as continuously variable transmissions that always find the most efficient gear, or engines that are designed to shut down one cylinder until it is needed.

Even before the final implementation plan had been prepared, other states followed California's lead. Connecticut, Maine, Massachusetts, New Jersey, New York, Rhode Island, and Vermont were considering measures similar to the California plan.

Automakers disagreed with CARB's cost projections, arguing that manufacturing costs would be higher. Near the end of 2004, they sued the state, claiming that California was usurping federal jurisdiction by seeking to control the gas mileage of vehicles (fig. 23). Alan C. Lloyd, the CARB chairman, made clear his frustration with the auto industry when he told the *Los Angeles Times* on September 24, 2004 (after the Board adopted the new regulations): "'We sent letters to all the auto companies in the world,' asking for cooperation. 'The response, the silence, was deafening'" (Bastillo 2004).

California still has a long way to go to achieve its clean air goals; nearly all Californians still breathe unhealthy air at times.

—CARB (2001, 1)

Who Takes Regulatory Responsibility?

Air quality is regulated through both federal and state requirements and standards, plus regulations set by local air districts. The U.S. Environmental Protection Agency sets ambient air quality standards to protect public health and welfare, and it oversees state programs. The EPA also provides technical and financial assistance through the Office of Air and Radiation, the Office of Transportation and Air Quality, and the National Vehicle and Fuel Emissions Laboratory.

The EPA has the exclusive authority to regulate interstate trucks registered outside California, as well as certain new farm and construction equipment, new locomotives, ships, and aircraft. In most states other than California, it has *sole* jurisdiction over *all* cars and trucks, but an exception was made for this state, recognizing its unique need for more stringent controls, when the Federal Air Quality Act of 1967 was passed. Since 1990, other states have been allowed to adopt the California program, and several have done so.

When areas within states are in "nonattainment" status for federal standards, a State Implementation Plan (SIP) is required. The federal government can apply sanctions for failure to plan or to meet deadlines for implementing an SIP. Federal highway funds might be withheld, for example. In 1994, because there were numerous nonattainment areas in California, a federal court ordered the EPA to develop a Federal Implementation Plan (FIP) for California. However, the state "SIPed the FIP" by submitting an acceptable state plan.

The California Air Resources Board (CARB), a department within the California Environmental Protection Agency, includes an 11-member board appointed by the governor and several hundred staff members. CARB's mission is to protect the public health, welfare, and ecological resources in California by reducing air pollution. The agency is also required to consider the impact on the state's economy when it sets regulations. The state board oversees the regulatory activity of California's 35 local air districts.

Responsibility for pollution from stationary sources, like factories or businesses, lies with local air districts (map 16). County Air Pollution Control Districts and regional Air Quality Management Districts, which are governed by local elected officials, develop local attainment plans, issue permits to regulate stationary sources, handle smoke management and decisions about burn days, and approve local or regional transportation control measures. Under the California Clean Air Act, local air districts in nonattainment status must prepare plans to attain standards by the earliest practicable date.

Statewide, CARB supervises a network of more than 200 air-monitoring stations. It maintains an emissions inventory that identifies where problems exist for each type of pollutant. Its scientists develop atmospheric models, and the agency sponsors health effects studies.

Tensions over Jurisdiction

Manufacturers whose products must be designed to comply with standards benefit from the cost-effectiveness of meeting just one set of nationwide requirements. Clearly, it would be impractical to allow many different states and localities to set different standards. On the other hand, the waiver for California recognized the state's special challenges and the economic reality of California's vast automobile and fuel markets. Cars, trucks, buses, and SUVs built for sale in California have to meet stricter emissions standards. Gasoline and diesel fuels also must be formulated to comply with state specifications. And although California's laws are generally more protective than the federal government's, manufacturers and retailers still make handsome profits in the state.

California's requirements spurred innovation in emissions-reduction technologies. For decades, the "trickle up" approach to air quality control, with California in the lead, has benefited

Map 16. California air districts.

1. Lake
2. Colusa
3. Feather River
4. Yolo-Solano
5. Sacramento Metro
6. Amador-Calaveras
7. Mariposa
8. Antelope Valley
9. Ventura
10. Northern Sonoma

the whole nation. Not every federal administration, however, has shared the same philosophy about federal versus state control. In several recent cases, the federal government supported lawsuits brought by manufacturers against specific regulatory actions in California.

Automakers secured administration backing in a 2002 lawsuit that challenged California's Zero Emissions Vehicles (ZEV) law. The manufacturers argued that California had preempted federal jurisdiction over fuel economy standards because the new law defined hybrid vehicles, in part, based on their miles-per-gallon ratings.

In a case before the U.S. Supreme Court, *Engine Manufacturers Association v. South Coast Air Quality Management District,* diesel engine manufacturers opposed a South Coast Air Quality Management District rule that required cities and owners of vehicle fleets to replace worn-out diesel-burning vehicles with clean-fuel versions. The engine makers asserted that the local standards would improperly preempt national ones. The Bush administration again sided with the industry, and the court agreed in 2004.

It is understandable why legal tests of the state's new Automotive Greenhouse Gas law make its fate uncertain.

Cleaning the Car

Because the majority of the state's air pollution originates on its highways, many of CARB's most important programs target pollution from cars and trucks (pl. 79). The state began removing lead from gasoline in 1976. Reformulations for cleaner-burning gasoline and cleaner diesel fuels have included lowering the content of volatile components and adding oxygenates to cut emissions. By 1999, it became clear that one additive, MTBE, was a major source of water pollution and a health hazard. It was gradually removed from California fuels, and ethanol (a renewable source made from corn) was substituted. However, CARB has requested—and received—several temporary waivers from the EPA because the board does not believe that California's gasoline actually needs ethanol to meet air quality standards. Pressure to use ethanol comes from the agricultural industry. Many view this debate as an economic one, more than an environmental issue.

Plate 79. Rush-hour freeway traffic passes refineries in Benicia.

Smog Check Program

The statewide Smog Check Program is not administered by CARB, but by the Bureau of Automotive Repairs in the California Department of Consumer Affairs. Periodic smog checks are designed to identify vehicles that need maintenance or repairs to their emission control systems. The program was enacted in 1979, took effect in 1984, and has been strengthened several times. In 1994, Smog Check II targeted vehicles that pollute at least two to 25 times more than average; these offenders had to be repaired and retested.

For vehicles in most parts of the state, smog checks are required every other year. Car owners receive reminders with their vehicles' registration renewal notices. A smog check also must be done when a vehicle is sold or when it is registered for the first time in California. Sellers must provide buyers with proof that the car passed a smog check inspection within the preceding 90 days.

In many rural parts of the state, where the air is cleaner, no inspection is required until a vehicle is sold (or when registering one for the first time within California). And smog checks are not mandatory for all types of vehicles in California. State law exempts diesel-powered vehicles, hybrid-electric vehicles, motorcycles, electric vehicles, and vehicles with two-cycle engines. Newer vehicles don't have to begin their every-other-year smog checks until they are six years old (or change owners during that period). An exemption for vehicles 30 years old or more was repealed in 2005. Beginning with the 1976 models, all vehicles must now be smog-checked.

Getting Older Cars off the Road

About 5 percent of the state's vehicles are responsible for almost half of its automotive pollution. Most of the dirtiest cars, trucks, and buses are older models. In California, you now can be paid to stop driving an old "junker." The Voluntary Accelerated Vehicle Retirement Program takes cars off the road in order to bring an earlier end to their years of polluting. Eligible vehicles must be operational and must have been registered for at least two years in the district where they are being retired. The car cannot be falling to pieces or nonoperational; the whole idea of the program is to shorten the life of an operating car that pollutes more than most.

How much might such a car be worth? The value depends on the air district where your vehicle is registered, but typically translates into a payment of $400 to $700.

As of 2006, this program was only available in the South Coast Air Quality Management District and the Bay Area Air Quality Management District, but there were plans to expand it.

Reporting Smoking Vehicles

A CARB study found that motorists traveling directly behind a smoking vehicle can be exposed to very high levels of pollution. Pollution levels inside a vehicle may be up to 10 times higher than in the outside air. As much as one-half the pollutants inside a car can come from a vehicle traveling directly ahead of it.

The program for reporting smoking vehicles asks callers (who telephone 1-800-END-SMOG) to provide a vehicle license number, its make and model, and the time and location where it was seen. The registered owner will not be fined or go to jail, but will receive a letter encouraging him or her to have the vehicle checked and repaired. While no other legal steps are taken, hopefully the shock of being reported will jar the owner into action. CARB estimates that half of those contacted do actually make repairs. Vehicles emitting excessive tailpipe exhaust are in violation of the California Vehicle Code, and owners of such vehicles can be stopped by law enforcement officers and receive citations.

Weaning Ourselves from Oil

What if exhaust pipes were not at the tail ends of our cars? What if they were positioned up front, where we would have to breathe our own exhaust poisons, instead of heedlessly driving away from them as we do now? That would be a crazy way to design cars, but it would certainly wake us up to some of the real air quality costs associated with burning fossil fuels.

What if cars and trucks could be designed to produce no pollution? At the same time, might we end our dependence on oil, with benefits that extend well beyond air quality? As fantastic as these questions may seem, in the near future such notions may become reality. "Political leaders with foresight would see that genuine energy independence comes not from adding a spoonful of Alaskan oil or a dollop of conservation . . . but from encouraging the speedy development of alternatives to oil-fired transport" (Vaitheeswaran 2003, 114).

CARB certifies types of low- and zero-emission vehicles (ZEVs). The "partial zero-emission vehicle" has near-zero evaporative emissions and comes with a 15-year/150,000-mile warranty on the emissions equipment. The "advanced technology partial zero-emission vehicles" make use of "ZEV-enabling" clean technologies such as alternative fuel or electric drives. The extensive lists and categories found at the

CARB website may seem daunting, but the good news is that so many low-emission car and truck models are available, certified to high standards of efficiency.

Hybrid vehicles with both gas- and electric-powered motors have become increasingly popular. In 2004, most dealerships had created waiting lists to serve the backlog of orders for these vehicles. California, as usual, led this transition. In 2003, there were 11,425 hybrids registered in this state—a quarter of the nation's total. In 2004, Sacramento became the first U.S. city with hybrid FedEx delivery trucks. They were designed to decrease particulate emissions by 90 percent, reduce smog-causing emissions by 75 percent, and travel 50 percent farther on a gallon of fuel.

With "full hybrids," an electric motor handles propulsion at low speeds typical of city driving, while the gasoline engine automatically kicks in at higher speeds. "Mild hybrids" use some gasoline all of the time. In both cases, electric batteries get recharged as the car travels and brakes, so they never need to be plugged into an electrical outlet. Hybrids can offer two to three times the energy efficiency of comparable gasoline-only cars, with ranges of about 600 miles on a tank of gas. Two automakers, Honda (with its Insight and Civic models) and Toyota (with its Prius), were the first to market hybrids, but other automakers are bringing models to the market soon, in sedan, SUV, and truck models.

Although hybrids bring improved mileage, they continue to rely on internal combustion engines and fossil fuels. Zero-emission vehicles take innovation to the next level. As of September 1999, there were more than 3,000 electric vehicles on the road in the nation. Most of them were in California. In 1990, in another example of "technology forcing" (such as the tactic used to push catalytic converters), California mandated that 10 percent of vehicles sold in the state in the 2003 to 2008 model years had to be zero-emission vehicles. The mandate was resisted by automakers and modified in 2001 to allow hybrid vehicles, which do emit some pollution, to qualify as part of the 10 percent requirement. The ZEV requirement may have stalled, but it did spur the development of hybrids.

Electric vehicles are the only true zero-emission vehicles on the road. Their batteries can be recharged at stations around the state. But there will still be significant emissions related to electric vehicles, because ultimately the electricity for recharging must come from "upstream" power plants.

Clean Diesel

Particulate pollution from diesel engines has been reduced with the introduction of biodiesel fuel made from recycled vegetable oil. A renewable resource, vegetable oil originates from plants. A standard diesel car or truck can be converted into a "veggie car," as they have been called. The engines in these cars must be started using regular diesel fuel, since vegetable oil is too thick until it becomes heated. Once the engine is warm, it can switch to using the biodiesel fuel only.

A new generation of high-performance "clean" diesels has been developed in Europe. The design does not perform well with the diesel fuel currently used in the United States, which contains a lot more sulfur (and therefore is also more polluting) than European diesel. New EPA rules that will lower the sulfur content of U.S. diesel to European levels will take effect in 2006. Similarly, lubricants are being reformulated for use in the new clean diesels.

Hydrogen Fuel Cells

If growth forecasts come true, even if every single vehicle in the nation is a hybrid by 2025, we will still need to import as much oil then as we import today. Promoting hybrids is a good transitional strategy, but not a permanent solution to our problems with oil.

Hydrogen fuel cells may offer that solution. Fuel cells are electrochemical devices that produce electricity more efficiently than conventional engines, silently and without combustion. Unlike batteries, which also make power through chemical reactions, fuel cells never wear out or need to be recharged. They produce electricity as long as there is fuel in a refillable storage tank.

Storing hydrogen is a challenge. Hydrogen is the smallest atom on Earth and hydrogen gas (H_2) is the smallest molecule. The tiny molecule can escape from almost all containers. Unless a good storage tank is provided, you might leave your fully fueled hydrogen car parked at an airport and return in a couple of weeks to an empty tank. But conventional designs have only managed to achieve sufficient capacity with bulky, heavy storage tanks. Designers are certain, though, that this problem can be resolved.

Fuel cell vehicles are 50 percent efficient, compared to perhaps 15 percent efficiency for gasoline combustion engines. That efficiency means that the cost of hydrogen fuel, per mile driven, ought to be significantly lower than fuel costs for conventional cars and also lower than gasoline costs for hybrids.

What about safety? Any fuel source contains energy and has to be handled properly. According to the California Fuel Cell Partnership (a public-private venture to demonstrate fuel cell vehicles in California), hydrogen is safer than the very volatile gasoline we now use. Those concerned about safety, however, never fail to mention the explosive Hindenburg blimp fire, early in the twentieth century. Yet, contrary to some misconceptions, what caused that fire was not the fuel in the hydrogen-filled airship, but the paint used on the blimp's exterior. The dramatic flames primarily burned the Hindenburg's framework.

Hydrogen can be produced by using energy to break apart water molecules. Currently, it is cheapest to liberate the hydrogen from fossil fuels. Natural gas is now the source for more than 90 percent of hydrogen gas used in fuel cells. While natural gas or methane will be the primary sources at first, in the long term hydrogen will be made from water using renewable energy like sunlight and wind power. At that point (perhaps decades away, unfortunately), car fuel will generate no smog-forming emissions or greenhouse gases, other than water.

CARB has forecast mass production of hydrogen fuel cell cars by 2014. Governor Schwarzenegger signed an executive order, in 2004, committing the state to develop a "hydrogen highway" refueling network with hundreds of stations (pl. 80).

Plate 80. The hydrogen fueling station at UC Davis, part of California's developing "hydrogen highway."

In the Palm Springs region, the SunLine Transit Agency has become a world leader by converting its bus fleet to hydrogen fuel. The agency uses solar energy to produce the hydrogen and also plans to use wind energy. Transit systems are a good way to introduce the public to this new technology. Buses can handle big fuel tanks until more compact fuel storage is successfully designed. They may become rolling "billboards" and rolling classrooms to help educate the general public about the changes ahead.

Renewable Energy, Here Today, Much More Tomorrow

California is the tenth-largest energy consumer in the world. Much can be done to reduce the amount of air pollution and greenhouse gases that are released as energy is generated. In 2002, California adopted the largest Renewable Energy Portfolio Standard in the country, requiring the state's electricity providers to double their use of renewable technologies such as wind and solar power by 2017. If all goes according to plan, by 2010 renewable energy sources will meet 20 percent of California's electricity needs.

Air Power

California alone boasts 60 percent of the nation's wind-generation capacity (map 17). In 2000, California produced 3.69 billion kilowatt hours with wind energy, but that was only 1.27 percent of the state's total electrical capacity. In 2002, there were five areas with turbines, taking advantage of the natural topographic regions where winds blow between air basins. That year, Altamont Pass, east of Livermore, had 4,788 turbines that generated enough energy to power almost 200,000 average households for a year (pl. 81). Farther south in the Coast Range, 168 turbines dotted Pacheco Pass. Near Palm Springs, San Gorgonio Pass was forested with 2,556 wind turbines. Solano captured energy from the delta breezes with 607 turbines, and along the crests of ridges lining Tehachapi Pass, winds from the San Joaquin Valley turned 3,445 turbines as they blew into the Mojave Desert.

Many turbines were installed back in the 1970s, when federal tax credits encouraged utilities to build wind energy systems. Not all of the wind farms built then were actually operated, though. After an administration change in Washington, incentives disappeared. In the 1990s, the industry awoke after new tax credits were tied to actual energy production.

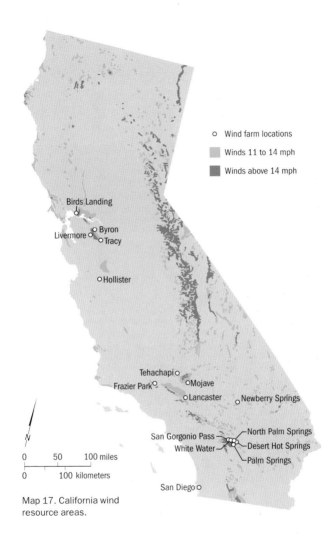

Map 17. California wind
resource areas.

The cost of wind-generated electricity is down to about
five cents per kilowatt-hour, which is comparable to the cost
of natural gas. The Energy Commission estimates that newer
technologies can reduce the cost of wind energy to 3.5 cents

Plate 81. Wind turbines at Altamont Pass wait to generate electricity on a windless day as fog fills the Central Valley.

per kilowatt-hour. Some current wind projects in other states have been bid at this lower cost.

Wind-generated electricity produces no air pollution but does come with environmental costs. Large clusters of wind turbines kill birds. A 2004 report to the California Energy Commission on reducing bird mortality at the Altamont Pass Wind Resource Area estimated that between 1,766 and 4,721 birds, including 881 to 1,300 raptors, die there each year. The Altamont Pass wind turbines kill 75 to 116 Golden Eagles (*Aquila chrysaetos*), 209 to 300 Red-tailed Hawks (*Buteo jamaicensis*), 73 to 333 American Kestrels (*Falco sparverius*), and 99 to 380 Burrowing Owls (*Athene cunicularia*) annually.

Reducing bird collisions with spinning blades is feasible, however. Old turbines should be replaced with fewer, larger models on taller towers, according to the report. Newer turbines can generate 10 times as much electricity as older ones, and their larger blades are easier for birds to see and avoid. Turbines installed in "wind walls," with parallel rows of closely aligned turbines (but with alternating heights) prove to be the safest configuration for birds.

In 2005, Altamont Pass wind turbine owners complied with the recommendations and also agreed to close half of the wind turbines in the pass for three months each winter. Alameda County then dropped its moratorium on permits for additional wind towers; the county had blocked expansion of the facility until the wind farm could demonstrate progress in reducing bird kills. A lawsuit against the electrical utilities, brought by the Center for Biological Diversity, was dropped after the report was published.

Solar

We have mostly ignored one of this state's most abundant resources: the sun. That may change. In 2006, the state's largest solar program ever was approved by the State Public Utilities Commission. Rebates (totaling $2.9 billion by 2016) aim to encourage the installation of solar panels on one million homes and businesses (pl. 82). The rebates will start at $2.80 per watt (about $7,000 for a standard household system), then gradually decrease over time, the assumption being that costs will decline with mass production.

Plate 82. Solar electric panels serving the author's home.

When energy is generated on-site, massive failures in the interstate power grid could become irrelevant. If we took full advantage of solar electric systems, demand for imported fossil fuels would drop significantly.

California now makes 56 percent of its electricity from coal and gas. Between 2004 and 2014, the state's energy demands may increase 20 percent, if one million new homes are constructed (as forecasts suggest). The California Energy Commission estimates that California will require 500 to 1,000 megawatts of new energy capacity each year to accommodate such growth. If those in power respond to these forecasts simply by building fossil-fuel power plants, millions of tons of additional pollution will enter California's air.

After California's 2001 energy crisis, demand for solar power jumped by a factor of 10. Most of that demand came from existing homes and businesses that took advantage of the state's Emerging Renewals Rebate Program. In 2004, rebates were set at $3.00 per watt for solar photovoltaic systems installed by qualified contractors.

With state funding, solar systems were added to buildings at a dozen of California's fairgrounds by 2004. The systems ranged from more than 9,000 panels in San Diego County down to 1,050 panels at the Colorado River Fairgrounds in Blythe. Another reaction to the rolling blackouts that hit the state in 2001, this program aimed to replace 25 percent to 75 percent of each fairground's electricity use. Money saved in utility costs was targeted to pay for the installations and fund the statewide conversion program. Eventually, 50 of the state's 74 fairgrounds should have solar electric systems.

Cool Communities

Los Angeles has 1,000 square miles of roof and roads that heat up on sunny days, creating an "urban heat island effect" that raises temperatures in that southern California basin by about 7 degrees F. Higher temperatures mean that more photochemical smog forms. Up on the roof, where dark material absorbs heat, peak temperatures may reach 150 to 190 degrees F on hot summer days. And everybody knows how an asphalt parking lot can bake in the sun. Light-colored materials that reflect more of the sun's energy can be 50 to 60 degrees cooler.

The Heat Island Group at the Lawrence Berkeley National Laboratory has found that buildings in Sacramento with lightly colored, more reflective roofs use up to 40 percent less energy than those with darker roofs. According to the Group, peak power demands rise 3 percent for every half-degree rise in daily maximum temperature. In Los Angeles, about 1 to 1.5 gigawatts of power annually must be used to cool everybody off and compensate for the "heat island" impact. This costs the state's electricity consumers about $100,000 per hour, totaling $100 million per year.

"Cool roofs" exist because NASA needed to develop reflective materials to keep spacecraft from overheating in the intense sunlight of space. The latest high-tech roofing materials, amazingly, come in the full range of traditional dark roofing colors, yet with layering that still reflects almost all of the sun's heat energy. Beginning in 2008, California law will require that every new roof be "cool." Local utilities like Pacific Gas & Electric and Southern California Edison offer rebate programs for installation of cool-roof materials.

Similar advantages occur when black asphalt is replaced by material the color of aged concrete; "cool pavements" even last longer than hot ones do. Cars with white or silver roofs can help keep the interiors of vehicles cool, reducing the fuel costs from running air conditioners.

Success Stories

State and federal agencies responsible for our air and energy resources have many success stories to point to in the last half-century. In this final example, California leadership again pioneered a change that spread across the nation.

In 1974, the average refrigerator used 2,200 kilowatt-hours of energy each year. Today, not only have replacements been mandated for ozone-depleting CFCs, but energy consumption by refrigerators is one-fourth what it was 30 years ago, averaging only about 600 kilowatt-hours. And prices of new refrigerators have actually fallen. The technology-forcing standards that led to the efficient refrigerator were mandated by the California Energy Commission. Today, the federal "Energy Star" designation helps consumers across the nation select the most efficient appliances and products.

Much progress has been made:

> The number of unhealthy days has improved considerably across the State, down by almost half between 1980 to 2000 in Los Angeles. The decline in statewide health risk from air toxics, like benzene and lead, has been equally dramatic. (CARB 2001, 1)

Yet governments can only do so much. Not all of our air quality problems originate with businesses, factories, or the decisions reached (or avoided) by politicians. We should recognize the impacts of our own choices, multiplied by millions, and take individual responsibility too.

What *You* Can Do, Every Day

Every time you drive, turn on an air conditioner, mow your lawn, or paint a room, you make choices that affect air quality. Seemingly insignificant decisions, such as using water-based instead of oil-based paints, can reduce air pollution. Oil-based paints contain three to five times more volatile smog-forming compounds than water-based paints.

There are probably an endless number of differences you can make every day. When overwhelmed, keep in mind three overriding goals: be a thoughtful consumer, never waste energy, and never tolerate dirty air. Our adaptability as a species can produce unfortunate complaisance. Atmospheric scientist Richard Turco has noted, "We are a society of persons so out of touch with nature that crowded freeways and gasoline fumes seem natural" (1997, 180). But clean air and a healthy global atmosphere are your birthrights. You can translate those goals into daily actions, as you'll see in the following sections.

Cars

Never top off a gas tank. Invariably, you will spill some gas and catch a whiff of fumes. Those signs on the pump are trying to save you from unnecessary exposure to carcinogenic benzene, after all.

Maximize your gas mileage by keeping your vehicle tuned up, changing the air filter at recommended intervals, and keeping the tires properly inflated. Going easy on the gas pedal will help too; avoid fast accelerations. Cruise control can help on the freeway. Most Californians have more than one vehicle available to them. Use the one that is most fuel-efficient, usually the newest car, for most of your trips.

Buy a car that gets really good gas mileage. Seek out a low-emission, super or "ultra" low-emission, or hybrid gas-electric vehicle. Watch for (and support) opportunities to join California's "hydrogen highway."

Take the bus, carpool, or vanpool. Californians really can do such things, despite our well-known attachment to cars. Or, do something not just un-Californian, but unlike most Americans: bicycle instead of drive. Get some exercise, generate *no* air pollution, save money on gas, and get in shape.

At Home

Draw the blinds when it is hot. Close your windows during the day and open them at night when it is cool. In colder climate regions, open blinds during the daytime to allow sunlight to heat your home and close them at night to help insulate from heat loss through window glass.

Lower your heater thermostat settings and raise them on your air conditioner. Programmable thermostats can automatically turn off the air conditioner or heater when you are not home or are sleeping. Or simply use a fan instead of air conditioning.

Turn off the lights when you leave a room. Replace incandescent light bulbs with energy-efficient, long-lasting compact fluorescents. They cost more up front but last much longer and save money on both electric bills and replacement costs. Always look for the most energy-efficient appliances, with "Energy Star" labels. Less efficient models may be on sale, but they will cost more to operate.

Caulk and weather-strip around doors and windows. Add insulation to your attic. When replacing windows, choose the best energy-saving versions.

Wait to run the dishwasher until it is truly full. Use the energy-saving setting to let dishes dry naturally. Or simply wash dishes by hand. Set the washing machine to clean clothes in warm or cold water. Turn down the thermostat on your water heater to no more than 120 degrees. Or consider replacing it with an on-demand hot water system. Why heat all that water in the tank while you are at work, when you are asleep at night, or even when you go away on vacations?

Mow down air pollution by converting to an electric lawnmower or push mower. A gas mower that's run for one hour can emit the same amount of VOCs as a car driven 50 miles. See if your local air district sponsors a replacement program. The South Coast Air Quality Management District took in over 4,000 old gas mowers in 2004 and, for $100, replaced them with electric mowers with the same power as a five-horsepower gas mower. That exchange reduced hydrocarbon emissions by nearly 20 tons a year, equivalent to the emissions from 43 new cars driven 12,000 miles each.

Fully participate in your local recycling programs. Look for opportunities to buy products made from or packaged in recycled material. If it means paying slightly more, consider it a good investment that can ensure that such programs thrive and grow. You will reduce air pollution from manufacturing processes and from landfills, incinerators, and trash-hauling trucks. The goal, as established by state law, is to recycle at least 50 percent.

What about Indoor Air?

Air filter systems for indoor air are popular but not always effective. They come in a number of design types. Mechanical filters force air through mesh filters to trap pollen, dust, and other particles. Electronic filters use charges to attract particles, which then stick to plates or other surfaces that must be cleaned. Gas-phase filters remove fumes and odors but do not work on particles. There are also ozone generators. Be cautious about this last option. Some manufacturers of ozone generators have labeled ozone "super oxygen," "mountain fresh air," or "activated oxygen" to disguise its toxic character. The EPA has analyzed the effectiveness of these machines and concluded that concentrations produced by ozone generators can exceed health standards even if you follow the manufacturer's instructions. It is difficult to control the amount of ozone produced relative to the size of your rooms and to prevent ozone from reacting with other materials in a room; ventilation variables present additional challenges. Ozone is generally ineffective in controlling indoor air pollution at concentrations below health standards.

One good alternative is a house plant. Or many house plants. The Associated Landscape Contractors of America and NASA studied common house plants to see which worked best at correcting the "sick building syndrome." The plants absorb contaminants through their leaves and roots, and bacteria in the potting soil also help. The top 10 plants for removing formaldehyde, benzene, and carbon monoxide were: bamboo palm *(Chamaedorea seifrizii);* Chinese evergreen *(Aglaonema modestum);* English ivy *(Hedera helix);* Gerbera daisy *(Gerbera jamesonii);* dracena Janet Craig *(Dracaena deremensis* 'Janet

Craig'); dracena marginata *(D. marginata);* mass cane/corn plant *(D. fragrans* 'Massangeana'); Mother-in-law's tongue *(Sansevieria laurentii);* peace lily *(Spathiphyllum* 'Mauna Loa'); pot mum *(Chrysanthemum morifolium);* and dracena 'Warneckii' *(Dracaena deremensis* 'Warneckii').

Finally, Think about Numbers

Do you know how many people live in California? A survey by the Public Policy Institute of California found that few Californians knew their state's population was more than 35 million in 2004, or that the numbers were forecast to reach 55 million by the year 2050 (map 18).

Population estimates are not certainties. We do not have to go there. Every time you read some statement suggesting that before long there "will be" 12 million more motor vehicles on our highway, for example, you should look for the qualifier: "if present rates of growth continue."

California's air quality challenges are exacerbated by our numbers. Since there are so many people in air basins overloaded with their emissions, perhaps we should refer to smog as "people fumes." That might keep us focused on this other, overriding challenge: someday the state's population *must* stabilize.

Then we all might breathe a little easier.

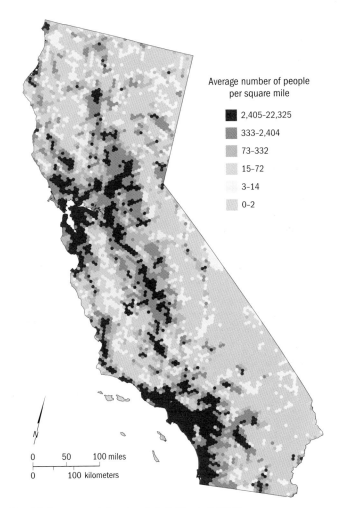

Average number of people
per square mile

- 2,405–22,325
- 333–2,404
- 73–332
- 15–72
- 3–14
- 0–2

0   50   100 miles

0   100 kilometers

Map 18. Human population density in California in 2002.

GLOSSARY

Aerosol Particles of solid or liquid matter that can remain suspended in air for a period of time from a few minutes to many months depending on the particle size and weight.

Air basin A land area with generally similar meteorological and geographic conditions throughout. To the extent possible, air basin boundaries are defined along political boundary lines and include both the source and receptor areas. California is currently divided into 15 air basins.

Air pollutants Amounts of foreign and/or natural substances occurring in the atmosphere that may result in adverse effects to humans, animals, vegetation, and/or materials.

Air Pollution Control District (APCD) A county agency with authority to regulate stationary, indirect, and area sources of air pollution, governed by a district air pollution control board composed of the elected county supervisors.

Air Quality Management District (AQMD) An agency that regulates stationary, indirect, and area sources of air pollution within a region and is governed by a regional air pollution control board comprised mostly of elected officials from within the region.

Air toxics Substances that are especially harmful to health, such as those considered under the EPA's hazardous air pollutant program or California's air toxics program.

Airshed A subset of air basin; a geographical area that shares the same air because of topography, meteorology, and climate.

Atmosphere The envelope of gases surrounding the Earth. From ground level up, the atmosphere is further subdivided into the troposphere, the stratosphere, the mesosphere, and the thermosphere.

Attainment area A geographical area that has air quality as good as, or better than, the national and/or California ambient air quality standards. An area may be an attainment area for one pollutant and a nonattainment area for others.

Climate The temperature, humidity, precipitation, winds, and other weather conditions over an extended period of time. (Compare to "weather.")

Coriolis force The deflection of winds and ocean currents due to the Earth's rotation. Deflection acts to the right in the Northern Hemisphere.

Criteria air pollutant An air pollutant for which acceptable levels of exposure can be determined and for which an ambient air quality standard has been set.

Emission standard The maximum amount of a pollutant that is allowed to be discharged from a polluting source such as an automobile or smoke stack.

Federal Implementation Plan (FIP) In the absence of an approved State Implementation Plan (SIP), a plan prepared by the EPA which provides measures that nonattainment areas must take to meet the requirements of the Federal Clean Air Act.

Fossil fuels Fuels such as coal, oil, and natural gas that are the remains of ancient plant and animal life.

Greenhouse gases Atmospheric gases (such as water vapor, carbon dioxide, methane, chlorofluorocarbons, nitrous oxide, and ozone) that slow the passage of re-radiated heat through the Earth's atmosphere.

Haze Reduced visibility due to the scattering of light caused by aerosols. Haze is caused in large part by man-made air pollutants.

Inversion A layer of warm air in the atmosphere that prevents the rise of cooling air and traps pollutants beneath it.

Limiting nutrient The nutrient that becomes exhausted first, thus limiting cellular growth. Nitrogen is often the limiting nutrient for plants.

Methyl tertiary butyl ether (MTBE) An ether compound added to gasoline to provide oxygen and enhance complete combustion. MTBE has been removed from California's gasoline.

Mobile sources Sources of air pollution that move around, such as automobiles, motorcycles, trucks, off-road vehicles, boats, and airplanes.

Nitrogen oxides (NO_x) Air pollution compounds of nitric oxide (NO), nitrogen dioxide (NO_2), and other oxides of nitrogen that may be created during combustion processes and are major contributors to smog formation and acid deposition.

Nonattainment area A geographic area that does not meet either national or state standards for a given pollutant.

Particulate matter (PM) Any material, except pure water, that exists in the solid or liquid state in the atmosphere. The size of particulate matter can vary from coarse, wind-blown dust particles to fine-particle combustion products.

Relative humidity The amount of water actually in the air compared to how much water the air can hold at a given temperature.

Sequestration The removal of a substance from one regime and its storage in another, as when carbon dioxide is sequestered from the atmosphere by plants doing photosynthesis.

Smog A combination of smoke and other particulates, ozone, hydrocarbons, nitrogen oxides, and other chemically reactive compounds that, under certain conditions of weather and sunlight, may result in a murky brown haze that causes adverse health effects.

Soot Very fine carbon particles that have a black appearance when emitted into the air.

State Implementation Plan (SIP) A plan prepared by states and submitted to the EPA describing how each area will attain and maintain national ambient air quality standards.

Stationary sources Nonmobile sources of pollution, such as power plants, refineries, and manufacturing facilities that emit air pollutants.

Toxic air contaminant (TAC) An air pollutant that may cause or contribute to an increase in deaths or in serious illness, or that may pose a present or potential hazard to human health.

Ultraviolet radiation (UV) The light energy range just beyond violet in the visible spectrum.

Virga Streaks or wisps of precipitation falling from a cloud but evaporating before reaching the ground.

Volatile organic compounds (VOCs) Hydrocarbon compounds that evaporate into the air. VOCs contribute to smog formation or may themselves be toxic. VOCs often have an odor; some examples include gasoline, alcohol, and the solvents used in paints. Also referred to as "reactive organic gases."

Weather The state of the atmosphere at a particular place and time (its temperature, cloudiness, wind, humidity, and precipitation).

REFERENCES

American Geophysical Union. 2003. *Human impacts on climate.* Policy statement. Washington, DC: American Geophysical Union. www.agu.org/sci_soc/policy/climate_change_position.html (accessed March 2006).

American Lung Association. 2004. *State of the air: 2004.* New York: American Lung Association. www.lungusa.org/site/pp.asp?c=dv LUK9O0E&b=50752 (accessed March 2006).

Anair, Don, and Patricia Monahan. 2004. *Sick of soot: Reducing the health impacts of diesel pollution in California.* Berkeley, CA: Union of Concerned Scientists. www.ucsusa.org/clean_vehicles/ big_rig_cleanup/sick-of-soot-solutions-to-californias-diesel-pollution.html (accessed March 2006).

Audesirk, Teresa, and Gerald. 1996. *Biology: Life on Earth.* Upper Saddle River, NJ: Prentice Hall.

Bach, Richard. 1970. *Jonathan Livingston Seagull.* New York: Avon.

————. 1974. *A gift of wings.* New York: Dell.

Badè, William Frederic. 1923. *The life and letters of John Muir,* vol. 1. Boston: Houghton Mifflin.

Bailey, Diane, Thomas Plenys, Gina M. Solomon, Todd R. Campbell, Gail Ruderman Feuer, Julie Masters, and Bella Tonkonogy. 2004. *Harboring pollution: The dirty truth about U.S. ports.* New York: Natural Resources Defense Council and Coalition for Clean Air. www.nrdc.org/air/pollution/ports/ports.pdf (accessed March 2006).

Bailey, Harry P. 1975. *Weather of southern California.* Berkeley and Los Angeles: University of California Press.

Bakker, Elna. 1971. *An island called California.* Berkeley and Los Angeles: University of California Press.

Baldassare, Mark. 2004. *PPIC statewide survey: Special survey on Californians and the future.* San Francisco: Public Policy Institute of California.

Balzar, John. 2004. It's the air up there. *Los Angeles Times,* June 5, E1.

Bastillo, Miguel. 2004. Tough car emissions rules OK'd. *Los Angeles Times,* September 25, B1.

Benedick, Richard Elliot. 1991. *Ozone diplomacy.* Cambridge, MA: Harvard University Press.

Berman, Daniel M., and John T. O'Connor. 1996. *Who owns the sun?* White River Junction, VT: Chelsea Green Publishing.

Brown, Bolton Coit. 1896. A trip about the headwaters of the South and Middle Forks of Kings River. *Sierra Club Bulletin* May, 1(8).

Bush, Mark B. 2000. *Ecology of a changing planet.* Upper Saddle River, NJ: Prentice Hall.

California Air Resources Board. 2001. *Ozone transport: 2001 review.* Summary report. Sacramento, CA: California Air Resources Board. www.arb.ca.gov/aqd/transport/assessments/assessments .htm (accessed March 2006).

————. 2003a. Appendix D: Surface area, population, and average daily vehicle miles traveled. In *The 2003 California almanac of emissions and air quality.* Sacramento: California Air Resources Board. www.arb.ca.gov/aqd/almanac/almanac03/almanac03 .htm (accessed March 2006).

————. 2003b. *Fifty things you can do.* Sacramento: California Air Resources Board. www.arb.ca.gov/html/brochure/50things.htm (accessed March 2006).

————. 2003c. *The ozone weekend effect in California.* Staff report. Sacramento: California Air Resources Board. www.arb.ca.gov/ aqd/weekendeffect/we_sr-final.htm (accessed March 2006).

————. 2004a. *California's air quality history key events.* Sacramento: California Air Resources Board. www.arb.ca.gov/html/brochure/ history.htm (accessed March 2006).

————. 2004b. *Fresno Asthmatic Children's Environment Study (FACES).* Sacramento: California Air Resources Board. www.arb .ca.gov/research/faces/faces.htm (accessed March 2006).

————. 2004c. *Maximum feasible and cost-effective reduction of greenhouse gas emissions from motor vehicles.* Staff proposal. Sacramento: California Air Resources Board. www.arb.ca.gov/cc/factsheets/cc_isor.pdf (accessed March 2006).

————. 2004d. *California air districts.* www.arb.ca.gov/ei/maps/ statemap/dismap.htm. (accessed March 2006).

————. 2005. *California air basins.* Sacramento: California Air Resources Board. www.arb.ca.gov/ei/maps/statemap/abmap.htm (accessed March 2006).

California Department of Finance. 2004. *Population projections by race/ethnicity for California and its counties, 2000–2050.* Sacramento: California Department of Finance. www.dof.ca.gov/ html/Demograp/DRU_Publications/Projections/P1.htm (accessed March 2006).

California Department of Fish and Game. 2003. *Atlas of the biodiversity of California.* Sacramento, CA: California Resources Agency.

California Department of Forestry and Fire Protection. 2000. *Maps: City lights.* http://frap.cdf.ca.gov/data/frapgismaps/select.asp (accessed March 2006).

California Department of Water Resources. 1985. *The California wind atlas.* Sacramento: California Energy Commission.

California Energy Commission. 2004. *West Coast Governors Global Warming Initiative.* www.climatechange.ca.gov/westcoast/documents/index.html (accessed March 2006).

———. 2006. *California wind resource maps.* www.energy.ca.gov/maps/wind.html (accessed March 2006).

———. n.d. *Climate change and California.* www.climatechange.ca.gov (accessed March 2006).

———. n.d. *Cool savings with cool roofs.* www.consumerenergycenter.org/coolroof (accessed March 2006).

California Legislature. 1999. *Climate change: Its economic and environmental effects.* Transcript of interim hearing, November 16. Senate Committee on Environmental Quality, Byron D. Sher, Chair. Los Angeles: California Legislature.

CARB. *See* California Air Resources Board.

Center for Clean Air Policy. 2006. *Cost-effective GHG mitigation measures for California.* Summary report. Washington, DC: Center for Clean Air Policy. www.ccap.org/domestic/Summary%20Report-Final%20_1-19-06_.pdf (accessed March 2006).

Chandler, Raymond. 1946. *Red wind.* Cleveland, OH: World Publishing.

Chase, J. Smeaton. 1913. *California Coast Trails.* Boston: Houghton Mifflin.

Chea, Terence. 2003. Californians worry about air pollution, but don't blame themselves. *Bakersfield Californian,* July 10.

Christianson, Gale E. 2000. *Greenhouse: The 200-year story of global warming,* reissue ed. New York: Penguin.

Coffel, Steve, and Karyn Feiden. 1991. *Indoor pollution.* New York: Random House.

Cosby, Bill. 1965. *Why is there air?* Record album. Burbank, CA: Warner Brothers.

da Vinci, Leonardo. 1956 [1508]. *The notebooks of Leonardo da Vinci—arranged, rendered into English, and introduced by Edward MacCurdy.* New York: George Braziller.

Davidson, Carlos, H. Bradley Shaffer, and Mark R. Jennings. 2001. Declines of the California Red-legged Frog: Climate, UV-B, habitat, and pesticides hypotheses. *Ecological Applications* 11 (2): 464–479.

———. 2002. Spatial tests of the pesticide drift, habitat destruction, UV-B, and climate-change hypotheses for California amphibian declines. *Conservation Biology* 16 (6): 1588–1601.

DeBlieu, Jan. 1998. *Wind: How the flow of air has shaped life, myth, and the land.* New York: Houghton Mifflin.

Doyle, Jack. 2000. *Taken for a ride: Detroit's big three and the politics of pollution.* New York: Four Walls Eight Windows.

Engle, Diana L., and John M. Melack. 1997. *Assessing the potential impact of acid deposition on high altitude aquatic ecosystems in California, integrating ten years of investigation.* Sacramento: California Air Resources Board. www.arb.ca.gov/research/abstracts/93-312.htm#Abstract (accessed March 2006).

Environmental Working Group. 1997. *People of color in California breathe the most heavily polluted air.* www.ewg.org/reports/caminority/caminority.html (accessed March 2006).

———. 1999. *What you don't know could hurt you: Pesticides in California's air.* http://www.ewg.org/reports/cadrift/pr.html (accessed March 2006).

Fawcett, Jeffry. 1990. *Political economy of smog in southern California,* vol. 15, *The environment, problems and solutions.* New York: Garland Publishing.

Feely, Richard A., Christopher L. Sabine, Kitack Lee, Will Berelson, Joanie Kleypas, Victoria J. Fabry, and Frank J. Millero. 2004. Impact of anthropogenic CO_2 on the $CaCO_3$ system in the oceans. *Science* 305:362–366.

Fiddler, Claude. 1995. *The High Sierra.* San Francisco: Chronicle Books.

Field, Christopher B., G. C. Daily, F. W. Davis, S. Gaines, P. A. Marson, J. Melack, and N. I. Miller. 1999. *Confronting climate change in California.* Cambridge, MA: Union of Concerned Scientists. www.ucsusa.org/global_warming/science/confronting-climate-change-in-california.html (accessed March 2006).

Gilliam, Harold. 2002. *Weather of the San Francisco Bay region.* Berkeley and Los Angeles: University of California Press.

Hall, Jane V. 1993. The atmosphere we breathe. In *California's threatened environment,* ed. Tim Palmer. Covelo, CA: Island Press.

Hardin, Garrett. 1968. The tragedy of the commons. *Science* 162: 1245–1248.

Hobbs, Peter V. 2000. *Introduction to atmospheric chemistry*. Cambridge, UK: Cambridge University Press.

Huning, James R. 1979. *Hot, dry, wet, cold, and windy: A weather primer for the national parks of the Sierra Nevada*. Three Rivers, CA: Sequoia Natural History Association.

Intergovernmental Panel on Climate Change. 2001. *Third assessment report—climate change 2001*. Summary for policy makers. Geneva: Intergovernmental Panel on Climate Change. www.ipcc .ch/pub/un/syreng/spm.pdf (accessed March 2006).

IPCC. *See* Intergovernmental Panel on Climate Change.

Joint Institute for the Study of the Atmosphere and Ocean. 2005. *The Pacific Decadal Oscillation (PDO)*. Seattle: Joint Institute for the Study of the Atmosphere and Ocean. www.jisao.washington.edu/ pdo (accessed March 2006).

Kals, W. S. 1982. *Your health, your moods, and the weather*. Garden City, NY: Doubleday.

Kelly, Walt. 1970. Pogo. Okefenokee Glen & Perloo.

Kurpius, Meredith R., and Allen H. Goldstein. 2003. Gas-phase chemistry dominates O_3 loss to A forest, implying a source of aerosols and hydroxyl radicals to the atmosphere. *Geophysical Research Letters* 30 (7): 1371.

Lappé, Marc, and Britt Bailey. 2002. *New considerations for evaluating pesticide impacts to endangered/threatened species*. Report to the Environmental Protection Agency. Case No. C00–3150 CW. Eureka, CA: Californians for Alternatives to Toxics. www.alterna tives2toxics.org/pdfs/MasterESAfinal.pdf (accessed March 2006).

Leggett, Jeremy. 2001. *The carbon war, global warming and the end of the oil era*. New York: Routledge.

Little, Charles E. 1995. *The dying of the trees*. New York: Penguin.

Lovelock, James. 1995. *The ages of Gaia,* updated and rev. ed. New York: W. W. Norton.

Marland, G., T. A. Boden, and R. J. Andres. 2003. Global, regional, and national CO_2 emissions. In *Trends: A compendium of data on global change*. Oak Ridge, TN: Carbon Dioxide Information Analysis Center, Oak Ridge National Laboratory, U.S. Department of Energy.

McClurg, Sue. 1998. The challenge of MTBE: Clean air vs. clean water? *Western Water,* July–August. www.water-ed.org/julyaug 98.asp (accessed March 2006).

Miller, Paul, and Joe McBride, eds. 1998. *Oxidant air pollution impacts* on the montane forests of southern California: A case study of the San Bernardino Mountains. Ecological Studies, vol. 134. New York: Springer-Verlag.

Muir, John. 1961 [1894]. The mountains of California. Garden City, NY: Anchor.

————. 1979 [1938]. John of the mountains. Madison: University of Wisconsin Press.

National Academy of Sciences. 2001. Climate change science: An analysis of some key questions. Washington, DC: National Academy Press. www.nap.edu/openbook/0309075742/html/1.html (accessed March 2006).

National Park Service. n.d. Sequoia and Kings Canyon National Parks: Air resources overview. www.nps.gov/seki/snrm/air/air.htm (accessed March 2006).

Nordhoff, Charles. 1873. California for health, pleasure, and residence. New York: Harper and Brothers.

Pearce, Fred. 2002. Global warming: A beginner's guide to our changing climate. New York: DK Publishing.

Peters, John M. 1997. Epidemiologic investigation to identify chronic effects of ambient air pollutants in southern California. Report to California Air Resources Board. Sacramento: California Air Resources Board. www.arb.ca.gov/research/abstracts/a033-186.htm (accessed March 2006).

Polakovic, Gary. 2003. U.S., state clash over environment. Los Angeles Times, September 14.

Pratt, Beth. 2003. The Sierra wave. Sierra Nature Notes 3 (January). www.yosemite.org/naturenotes/SierraWave.htm (accessed May 2006).

Sabine, Christopher L., Richard A. Feely, Nicolas Gruber, Robert M. Key, Kitack Lee, John L. Bullister, Rik Wanninkhof, C.S. Wong, Douglas W.R. Wallace, Bronte Tilbrook, Frank J. Millero, Tsung-Hung Peng, Alexander Kozyr, Tsueno Ono, and Aida F. Rios. 2004. The oceanic sink for anthropogenic CO_2. Science 305: 367–371.

Sagan, Carl. 1997. Pale blue dot. New York: Ballantine.

Santer, Benjamin D., R. Sausen, T.M.L. Wigley, James S. Boyle, Krishna AchutaRao, Charles Doutriaux, J.E. Hansen, G.A. Meehl, E. Roeckner, R. Ruedy, G. Schmidt, and Karl E. Taylor. 2003. Behavior of tropopause height and atmospheric temperature in models, reanalyses, and observations: Decadal changes. Journal of Geophysical Research 108(D1), doi 10.1029/2002JD002258.

SCAQMD. See South Coast Air Quality Management District.

Schoenherr, Allan A. 1992. *A natural history of California.* Berkeley and Los Angeles: University of California Press.

Schwartz, Peter, and Doug Randall. 2004. *An abrupt climate change scenario and its implications for United States national security.* Prepared for U.S. Department of Defense. Emeryville, CA: Global Business Network. www.gbn.com/ArticleDisplayServlet.srv?aid= 26231 (accessed March 2006).

Shnayerson, Michael. 1996. *The car that could.* New York: Random House.

Siegenthaler, Urs, Thomas F. Stucker, Eric Monnin, Diether Lüthi, Jakob Schwander, Bernard Stauffer, Dominique Raynaud, Jean-Marc Barnoia, Hubertus Fischer, Valerie Masson-Delmotte, and Jean Jouzel. 2005. Stable carbon cycle—climate relationship during the late Pleistocene. *Science* 310 (5752): 1313–1317.

Sierra Nevada Ecosystem Project. 1996. Air quality. In *Status of the Sierra Nevada,* vol. 1, ch. 9. Berkeley: University of California. http://ceres.ca.gov/snep/pubs/web/PDF/v1_ch09.pdf (accessed March 2006).

Smallwood, K. S., and C. G. Thelander. 2004. *Developing methods to reduce bird mortality in the Altamont Pass Wind Resource Area.* Final report. Sacramento: California Energy Commission. www.energy.ca.gov/pier/final_project_reports/500-04-052.html (accessed March 2006).

Somerville, Richard C. J. 1998. *The forgiving air: Understanding environmental change,* reprint ed. Berkeley and Los Angeles: University of California Press.

South Coast Air Quality Management District. 1996. *Electric vehicles and California's future prosperity.* Moraga, CA: Business/Technology Books.

———. 1997. *The Southland's war on smog: 50 years of progress toward clean air.* www.aqmd.gov/news1/archives/history/marchcov.html (accessed March 2006).

———. 2003. *Final 2003 air quality management plan.* Diamond Bar, CA.

———. n.d. *Dirty air: The health effects of air pollution.* www.aqmd .gov/forstudents/dirty_air.html (accessed March 2006).

Stevenson, Robert Louis. 1990 [1895]. The amateur emigrant. In *O California,* ed. Stephen Vincent. San Francisco: Bedford Arts.

Stoddard, John L., and James O. Sickman. 2002. *Episodic acidification of lakes in the Sierra Nevada.* Final report A132–048. Sacramento: California Air Resources Board.

Travis, D.J., A.M. Carleton, and R.G. Lauritsen. 2002. Contrails reduce daily temperature range. *Nature* 418: 601.

Trewartha, Glenn T. 1968. *An introduction to climate.* New York: McGraw-Hill.

Truman, Benjamin C. (Major). 1874. *Semi-tropical California: Its climate, healthfulness, productiveness, and scenery.* San Francisco: A.L. Bancroft.

Turco, Richard P. 1997. *Earth under siege, from air pollution to global change.* New York: Oxford University Press.

Twain, Mark. 1973 [1872]. *Roughing it.* Berkeley and Los Angeles: University of California Press.

Union of Concerned Scientists. 2004. *Climate change in California: Choosing our future.* Berkeley, CA: Union of Concerned Scientists. www.climatechoices.org (accessed March 2006).

United Nations Environment Programme. 2000. *The Montreal Protocol on substances that deplete the ozone layer as adjusted and/or amended in London 1990, Copenhagen 1992, Vienna 1995, Montreal 1997, Beijing 1999.* Nairobi, Kenya: United Nations Environment Programme. www.unep.org/ozone/Montreal-Protocol/ Montreal-Protocol2000.shtml (accessed March 2006).

———. n.d. *Vital climate graphics.* www.grida.no/climate/vital/ index.htm (accessed March 2006).

U.S. Environmental Protection Agency. 1999. *Smog: Who does it hurt?* Washington, DC: U.S. Environmental Protection Agency. www. epa.gov/airnow/health (accessed March 2006).

———. 2006. *1999 National-scale air toxics assessment: Estimated emissions, concentrations and risk.* www.epa.gov/ttn/atw/nata 1999/index.html (accessed March 2006).

———. n.d., a. *Air quality maps: Regional maps.* www.epa.gov/Region 9/air/maps/maps_top.html (accessed March 2006).

———. n.d., b. *EPA map of radon zones.* www.epa.gov/iaq/radon/ zonemap/California.htm (accessed April 2006).

———. n.d., c. *Good up high bad nearby.* www.epa.gov/oar/oaqps/ gooduphigh/ozone.html (accessed April 2006).

U.S. Geological Survey. 2000. *USGS research finds that contaminants may play an important role in California amphibian declines.* Press release. www.usgs.gov/newsroom/article.asp?ID=540 (accessed March 2006).

Vaitheeswaran, Vijay V. 2003. *Power to the people.* New York: Farrar, Straus and Giroux.

Warren, Earl. 1977. The memoirs of Earl Warren. Garden City, New York: Doubleday and Company.

Weber, John. 2000. Changing our weather one smokestack at a time. *Earth Observatory,* August 7. http://earthobservatory.nasa.gov/Study/Pollution/pollution.html (accessed March 2006).

Williams, Jack. 1997. *The weather book.* New York: Vintage.

Young, Louise B. 1977. *Earth's aura.* New York: Alfred A. Knopf.

Zabik, J.M., and J.N. Seiber. 1993. Atmospheric transport of organophosphate pesticide from California's Central Valley to the Sierra Nevada Mountains. *Journal of Environmental Quality* 22: 80–90.

ART CREDITS

Plates

All photographs are by the author except as noted below.

DEBBIE ALDRIDGE, University of California at Davis 80

FRANK BALTHIS 36, 41, 45, 54, 66, 79

CALIFORNIA DEPARTMENT OF WATER RESOURCES, Courtesy of 77

CALIFORNIA GOVERNOR'S OFFICE, Courtesy of 78

RYAN CARLE author photo

CLEET CARLTON 18

JACQUES DESCLOITRES, MODIS Rapid Response Team, NASA/GSFC 24

BRETT LEIGH DICKS 38, 57

DIVISION OF NATURAL RESOURCES, Sequoia and Kings Canyon National Parks, Courtesy of 73

EARTH SCIENCES AND IMAGE ANALYSIS LABORATORY, NASA Johnson Space Center, Courtesy of 3

KEN EATON 12, 16

WARREN GRETZ, DE/NREL 60, 81

RICK KATTELMANN 9, 31, 42, 45, 57, opener pp. 158–159

RUSS KERR, majestyofbirds.com 29

ED LINTON, DE/NREL 58

SCOT MARTIN, California State Parks, Courtesy of, 64

RICK MOORE, Grand Canyon Trust 65

LEW NUNNELLY 17

CAMILLE PARMESAN 76

JEFF SCHMALTZ, MODIS Rapid Response Team, NASA/GSFC 49

GREG SMITH 39

STEVE STERNER, SEAL Consultants (Goleta, California) 69

MERLIN D. TUTTLE, Bat Conservation International 32

UNIVERSITY OF CALIFORNIA LIBRARY, Department of Special Collections, Courtesy of 1

VANCE VREDENBERG 75

CHARLES WEBBER, California Academy of Science 46, 56

Figures

FIG. 1 Steve Greenberg, *Los Angeles Daily News,* copyright 1979. Reprinted with permission.

FIG. 3 Redrawn from Bush 2000.

FIGS. 4 and 5 Audesirk 1996.

FIG. 6 Redrawn from United Nations Environment Programme, n.d.

FIG. 9 Redrawn from Williams 1997.

FIG. 10 Redrawn from Gilliam 2002.

FIG. 11 Redrawn from Joint Institute for the Study of the Atmosphere and Ocean 2005.

FIG. 13 Redrawn from Williams 2002.

FIG. 18 Redrawn from California Air Resources Board 2003.

FIG. 19 Redrawn from U.S. Environmental Protection Agency, n.d., c.

FIGS. 20–22 Redrawn from United Nations Environment Programme, n.d.

FIG. 23 Steve Greenberg, Ventura County Star, copyright 2002. Reprinted with permission.

Maps

MAP 2 Redrawn from California Department of Fish and Game 2003.

MAP 3 Redrawn from Allan A. Schoenherr 1992.

MAP 4 Redrawn from California Air Resources Board 2005.

MAP 5 Redrawn from California Air Resources Board 2001.

MAP 6 Redrawn from California Air Resources Board 2001.

MAP 7 Redrawn from California Air Resources Board 2001.

MAP 8 Redrawn from California Air Resources Board 2001.

MAPS 9 and 10 Redrawn from U.S. Environmental Protection Agency n.d., a.

MAP 11 Redrawn from U.S. Environmental Protection Agency n.d., b.

MAP 12 Redrawn from Anair and Monahan 2004.

MAP 13 Redrawn from U.S. Environmental Protection Agency n.d., a.

MAP 14 Detail from Davidson et al. 2002.

MAP 15 Redrawn from California Department of Forestry and Fire Protection 2000.

MAP 16 Redrawn from California Air Resources Board 2004d.

MAP 17 Redrawn from California Energy Commission 2006.

MAP 18 Redrawn from California Department of Fish and Game 2003.

INDEX

ABOUT THE AUTHOR

David Carle received a bachelor's degree from the University of California at Davis in Wildlife and Fisheries Biology and a master's degree from California State University at Sacramento in Recreation and Parks Administration. He was a ranger for the California State Parks for 27 years. He worked at various sites, including the Mendocino Coast, Hearst Castle, the Auburn State Recreation Area, the State Indian Museum in Sacramento, and, from 1982 through 2000, the Mono Lake Tufa State Reserve. He has taught biology and natural history courses at Cerro Coso Community College (the Eastern Sierra College Center) in Mammoth Lakes. David has written several books, among them *Introduction to Water in California* (UC Press, 2004), *Water and the California Dream: Choices for the New Millennium* (Sierra Club Books, 2003), *Burning Questions: America's Fight with Nature's Fire* (Praeger, 2002), and *Mono Lake Viewpoint* (Artemisia Press, 1992).

Series Design:	Barbara Jellow
Design Enhancements:	Beth Hansen
Design Development:	Jane Tenenbaum
Illustrator:	Dartmouth Publishing
Composition:	Jane Tenenbaum
Indexer:	Thérèse Shere
Text:	9.5/12 Minion
Display:	Franklin Gothic Book and Demi
Printer and binder:	Golden Cup Printing Company Limited

Introduction to California Desert Wildflowers, Revised Edition, by Philip A. Munz, edited by Diane L. Renshaw and Phyllis M. Faber

Introduction to California Plant Life, Revised Edition, by Robert Ornduff, Phyllis M. Faber, and Todd Keeler-Wolf

Introduction to California Chaparral, by Ronald D. Quinn and Sterling C. Keeley, with line drawings by Marianne Wallace

Introduction to the Plant Life of Southern California: Coast to Foothills, by Philip W. Rundel and Robert Gustafson

Introduction to Horned Lizards of North America, by Wade C. Sherbrooke

Introduction to the California Condor, by Noel F. R. Snyder and Helen A. Snyder

Regional Guides

Sierra Nevada Natural History, Revised Edition, by Tracy I. Storer, Robert L. Usinger, and David Lukas